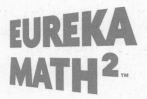

EUREKA
MATH²™

A Story of Units®

Units of Any Number ▸ 3

APPLY

Great Minds® is the creator of *Eureka Math*®,
Wit & Wisdom®, *Alexandria Plan*™, and *PhD Science*®.

Published by Great Minds PBC.
greatminds.org

Printed in the USA

1 2 3 4 5 6 7 8 9 10 QDG 26 25 24 23 22

ISBN 978-1-64497-650-0

Contents

Multiplication and Division with Units of 0, 1, 6, 7, 8, and 9

FAMILY MATH
Multiplication and Division Concepts with an Emphasis on Units of 6 and 8

Dear Family,

Your student is continuing to practice multiplication and division. They use their knowledge of multiplication and division with units of 3 and 4 to help them solve problems with units of 6 and 8. Familiar strategies, such as the break apart and distribute strategy, and familiar representations, such as arrays, equal groups, and tape diagrams, support your student as they multiply with larger factors. Your student is beginning to use a letter to represent an unknown value in models and equations.

6	$2 \times 3 = 1 \times 6$
12	$4 \times 3 = 2 \times 6$
18	$6 \times 3 = 3 \times 6$

Thinking about threes can help when multiplying and dividing with sixes because 6 is 2 groups of 3.

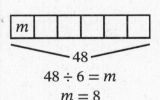

$$48 \div 6 = m$$
$$m = 8$$

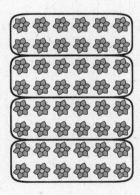

8	groups of (2 × 3)
8	× (2 × 3)
8	× 6

4	groups of (2 × 6)
4	× (2 × 6)
4	× 12

Letters can represent unknown values. It is useful to choose a letter related to the problem. For example, m could represent money.

Arrays can be broken into equal groups in different ways when they are being used to solve more challenging problems.

At-Home Activities

How Much and How Many with 6 and 8

With your student, look for opportunities to solve real-world multiplication and division problems with units of 6 and 8. For example, if a box of granola bars holds 6 bars, ask your student how many bars are in 5 boxes. If it costs $8 for a movie ticket, ask your student how many tickets you can buy with $24. Encourage them to use strategies such as skip-counting, drawing an array, and breaking apart the problem into more familiar factors.

Multiplication to Keep Fit

Do a fun workout with your student. Choose a few exercises, such as jumping jacks, sit-ups, push-ups, or lunges, that you can do together. Do each exercise 6 or 8 times to complete 1 set. Do 2 to 5 sets of the exercises with your student, keeping track of how many you complete. During your cooldown, ask your student questions about the number of exercises. Encourage them to use multiplication and division to determine the answers.

- "How many total sit-ups did we do? What can we multiply to help us figure it out?"
- "How many sit-ups and jumping jacks did we do? Tell me what equations help you find the total."
- "We did 48 lunges, and there were 8 in each set. How many sets of lunges did we do?"

1

Name _____

1. Carla sorts 15 toy cars into bins. She puts 5 toy cars into each bin.

   How many bins does Carla use?

   Write a division equation and a multiplication equation to describe the problem.

   Use a blank to represent the unknown.

   $15 \div 5 =$ _____

   $5 \times$ _____ $= 15$

I can use equal groups to help me think about the problem.

15 is the total.

5 is the size of each group.

The number of groups is the unknown.

I can write two equations to describe the problem.

$$\underline{\hspace{1.5em}15\hspace{1.5em}} \div \underline{\hspace{1.5em}5\hspace{1.5em}} = \underline{\hspace{2em}}$$

| Total | Size of each group | Number of groups |

$$\underline{\hspace{1.5em}5\hspace{1.5em}} \times \underline{\hspace{2em}} = \underline{\hspace{1.5em}15\hspace{1.5em}}$$

| Size of each group | Number of groups | Total |

Use the Read–Draw–Write process to solve the problem.

2. Ray and Pablo play a card game. Ray has 3 fewer cards than Pablo.

    Ray arranges his cards in 3 rows of 5 cards.

    How many cards does Pablo have?

    $3 \times 5 = 15$

    $15 + 3 = 18$

    Pablo has 18 cards.

I read the problem. I read again.

As I reread, I think about what I can draw.

I draw an array with 3 rows of 5 cards to represent Ray's cards.

I can multiply to find how many cards Ray has.

I know that Ray has 3 fewer cards than Pablo. I can also think of this as Pablo having 3 more cards than Ray.

Ray's Cards

| 15 | 3 |

?

Pablo's Cards

I can add to find the number of Pablo's cards.

Name _____

REMEMBER

1. Oka puts 35 muffins into baskets. Each basket has 5 muffins.

   How many baskets does she use?

   Write a division equation and a multiplication equation to describe the problem.

   Use a blank to represent the unknown.

Use the Read–Draw–Write process to solve the problem.

2. Shen and Liz set up chairs in a classroom. Liz makes 6 rows of 4 chairs.

   Shen says there are 2 fewer chairs than they need.

   How many chairs do they need?

Name _____

1. Each package of ice pops has 3 lime ice pops and 3 grape ice pops.

   a. Skip-count by threes to find the total number of ice pops.

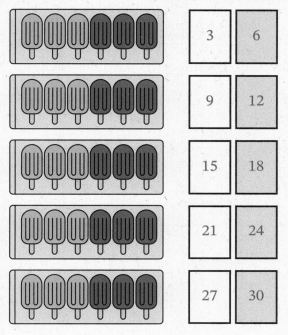

3	6
9	12
15	18
21	24
27	30

> There are 6 ice pops in each pack.
>
> I can skip-count by threes to help me skip-count by sixes.

   b. Complete the statements.

   10 threes is ___30___ .   5 sixes is ___30___ .

   ___10___ × 3 = ___30___   ___5___ × 6 = ___30___

> I see 5 packages of ice pops. Each pack has 2 groups of 3 ice pops.
>
> I know that skip-counting by threes 10 times has the same value as skip-counting by sixes 5 times.

   c. Use the pictures of ice pops to help you complete the statement.

   5 groups of 2 × ___3___ is the same amount as 5 × ___6___ .

2. Mia puts a total of 48 plums into 6 bowls. Each bowl has an equal number of plums. How many plums are in each bowl?

a. Draw and label a tape diagram that represents the problem. Label the unknown as *p*.

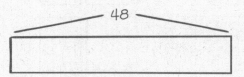

b. Write a division equation to represent the problem. Use the letter *p* for the unknown. Then find the value of *p*.

$48 \div 6 = p$

$p = 8$

I draw and label my tape diagram to represent all the plums.

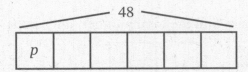

The plums have been put into 6 bowls, so I partition my tape diagram into 6 equal parts.

The unknown is the number of plums in each bowl. So, I label the unknown as *p*.

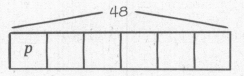

To help me solve, I can think: 6 groups of what number makes 48?

Name _____

1. Each bag of apples has 3 yellow apples and 3 green apples.

   a. Skip-count by threes to find the total number of apples.

   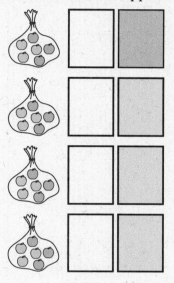

   b. Complete the statements.

   8 threes is _____.   4 sixes is _____.

   _____ × 3 = _____   _____ × 6 = _____

   c. Use the pictures of apples to help you complete the statement.

   4 groups of 2 × _____ is the same amount as 4 × _____.

2. Gabe puts a total of 36 donuts into 6 boxes. Each box has an equal number of donuts.

   How many donuts are in each box?

   a. Draw and label a tape diagram that represents the problem. Label the unknown as d.

   b. Write a division equation to represent the problem. Use the letter d for the unknown.

   Then find the value of d.

REMEMBER

Find the value of each unknown.

3. $2 \times 9 =$ _____

4. $24 \div 3 =$ _____

5. $50 \div 5 =$ _____

6. $7 \times 5 =$ _____

3

Name _____

1. Complete parts (a)–(c) to show the relationship between fours and eights.

   a. Skip-count by fours.

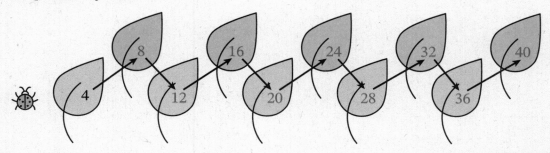

   b. Complete each statement to find the total.

   10 fours is ___40___ .   ___5___ eights is ___40___ .

   ___10___ × 4 = ___40___   ___5___ × ___8___ = ___40___

   c. Complete the statement to show the connection between the fours and eights.

   2 groups of 5 × ___4___ is the same amount as 5 × ___8___ .

   > I can skip-count by fours to help me skip-count by eights.
   >
   > I know that every other skip-count is a group of 8.
   >
   > 4, **8**, 12, **16**, 20, **24**, 28, **32**, 36, **40**

   > I know that skip-counting by fours 10 times is equal to skip-counting by eights 5 times.

Find the value of each unknown.

2. $p \times 8 = 48$

   $p = $ ___6___

3. $n \div 8 = 5$

   $n = $ ___40___

I think to myself that a number divided by 8 is 5. I use my skip-count and find what number I would say when I count 5 eights.

1 2 3 4 5
8, 16, 24, 32, (40,) 48, 56, 64, 72, 80

I think to myself that a number times 8 is 48. I skip-count by eights.

8, 16, 24, 32, 40, 48, 56, 64, 72, 80

I look at 48 and count the number of eights I skip-counted to get to 48.

1 2 3 4 5 6
8, 16, 24, 32, 40, (48,) 56, 64, 72, 80

---

4. Write an equation to represent the tape diagram. Then find the value of the unknown.

   $9 \times 8 = m$
   $m = 72$

8	8	8	8	8	8	8	8	8

   m

I see 9 groups of 8. The unknown total is m.

I can skip-count by eights to find the value of m.

REMEMBER

5. Shade 2 rows of 4.

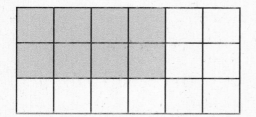

Fill in the number bond to represent the shaded rows of the array.

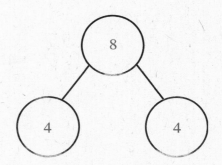

Write a repeated addition equation to find the total.

$$4 + 4 = 8$$

Rows are horizontal groups.
I shade 2 rows of 4.
I shade a total of 8 squares.

I write a repeated addition equation to match the shaded rows.

$$4 + 4 = 8$$

The value of each shaded row is four. Each part of my number bond is 4.

There are 8 tiles shaded. The total of my number bond is 8.

3

Name _____

1. Complete parts (a)–(c) to show the relationship between fours and eights.

   a. Skip-count by fours.

   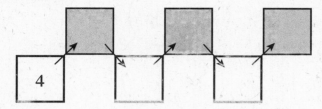

   b. Complete each statement to find the total.

   6 fours is _____.   3 eights is _____.

   _____ × 4 = _____   _____ × 8 = _____

   c. Complete the statement to show the connection between fours and eights.

   2 groups of 3 × _____ is the same as 3 × _____

Find the value of each unknown.

2. $p \times 8 = 40$

   $p =$ _____

3. $c \div 6 = 8$

   $c =$ _____

4. $8 \times w = 72$

   $w =$ _____

---

5. Write an equation to represent the tape diagram. Then find the value of the unknown.

   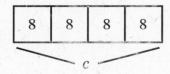

   Equation: _____

   _____

REMEMBER

6. Shade 3 rows of 2.

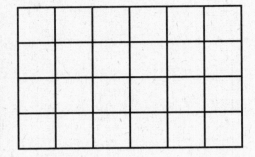

Fill in the number bond to represent the shaded rows of the array.

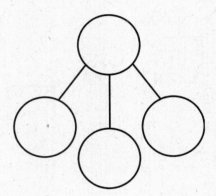

Write a repeated addition equation to find the total.

_____

4

Name _____

1. Use the array to help you fill in the blanks.

6×8

6 groups of (___2___ × ___4___)

$6 \times ($ ___2___ × ___4___ $)$

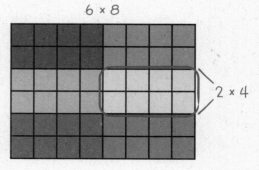

The large array has 6 rows of 8. There are smaller arrays that compose the large array.

6×8

2 × 4

I see 6 groups. Each group shows 2 rows of 4.

I can write this as 6 × (2 × 4).

Circle equal groups in the array. Then use the array to help you fill in the blanks.

2.

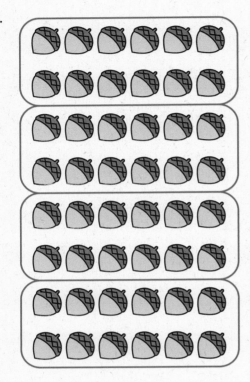

4 groups of (__2__ × __6__)

4 × (__2__ × __6__)

___4___ × __12__

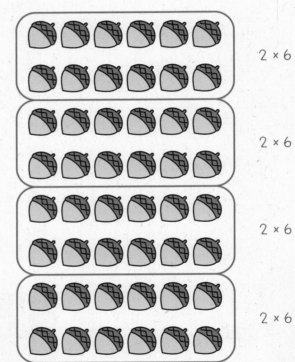

I can draw circles to show 4 equal groups.

Each group has 2 rows of 6 acorns.

I can write the number of acorns in each group as 2 × 6.

2 × 6

2 × 6

2 × 6

2 × 6

There are 4 groups of 2 × 6.

REMEMBER

3. Third grade students voted for their favorite fruit.

   The table shows how many students voted for each fruit.

   Use the data in the table to complete the scaled bar graph.

Type of Fruit	Number of Votes
Apple	25
Orange	15
Strawberry	50
Peach	30

Favorite Fruit

Number of Votes

Apple   Orange   Strawberry   Peach

Type of Fruit

> I need a scale that makes sense for the data in the table. The lowest value is 15. The highest value is 50.
>
> I can skip-count by fives. Each grid mark will represent 5 votes. I know I do not have to label every mark, so I only label the tens.

> Now I can draw bars for the data. The top of each bar lines up with a number on the scale.
>
> To graph the votes for apple, I draw a bar to 25 units on the scale.
>
> To graph the votes for orange, I draw a bar to 15 units on the scale.
>
> The bars to 15 and 25 stop halfway between labeled tick marks because they are halfway between multiples of 10.

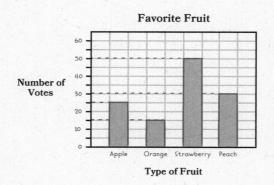

Name _____

Use the arrays to help you fill in the blanks.

1.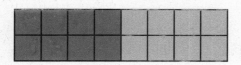

   2×8

   2 groups of (_____ × 4)

   $2 \times ($_____ × _____$)$

2.

   4×6

   4 groups of (_____ × _____)

   $4 \times ($_____ × _____$)$

Circle equal groups in each array. Then use the arrays to help you fill in the blanks.

3.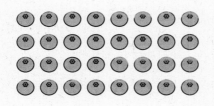

   4 groups of (_____ × _____)

   $4 \times ($_____ × _____$)$

   _____ × _____

4.

   3 groups of (_____ × _____)

   $3 \times ($_____ × _____$)$

   _____ × _____

REMEMBER

5. Third grade students voted for their favorite sandwich.

    The table shows how many students voted for each sandwich.

    Use the data in the table to complete the scaled bar graph.

Type of Sandwich	Number of Votes
Tuna	15
Peanut Butter	40
Turkey	55
Tomato	5

Favorite Sandwich

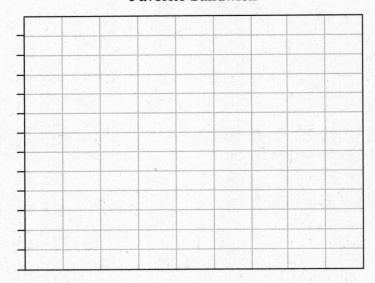

Number of Votes

Type of Sandwich

Name

Use the array to help you fill in the blanks and find the totals.

1.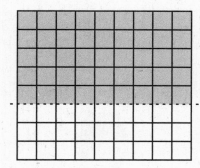

$5 \times 9 =$ ___45___

___3___ $\times 9 =$ ___27___

$8 \times 9 = (5 +$ ___3___ $) \times 9$

$= (5 \times 9) + ($ ___3___ $\times 9)$

$= 45 +$ ___27___

$=$ ___72___

The array shows 8 rows of 9. I can decompose the large array into smaller arrays to help me find the total.

I break apart the 8 rows into 5 rows and 3 rows. I can think of it as 5 nines + 3 nines.

I know 5 × 9 = 45 and 3 × 9 = 27.

I can add the two products to find 8 × 9.

Label the tape diagram. Then complete the equations.

2. $(5 \times 6) =$ __30__   __18__ $= ($ __3__ $\times 6)$

| 6 | | | | | | | |

$8 \times 6 = (5 + 3) \times 6$

$\qquad = (5 \times 6) + ($ __3__ $\times 6)$

$\qquad = 30 +$ __18__

$\qquad =$ __48__

> The tape diagram shows 8 sixes broken into 5 sixes and 3 sixes.
>
> 5 × 6 and 3 × 6 are familiar facts.
>
> I can find the product of 5 and 6 and add it to the product of 3 and 6.

---

3.   Shen reads 8 pages of his chapter book every day for 7 days.

How many total pages does he read?

Show how 8 × 7 can be broken apart into smaller facts to find the product.

Sample:

$8 \times 7 = (5 + 3) \times 7$

$\qquad = (5 \times 7) + (3 \times 7)$

$\qquad = 35 + 21$

$\qquad = 56$

> 8 sevens is the same amount as 5 sevens + 3 sevens.
>
> I know that 5 sevens is 35 and 3 sevens is 21.
>
> I can add 35 and 21 to find the total number of pages Shen reads.

REMEMBER

Third grade students voted for their favorite outside activity.

The scaled bar graph shows how many students voted for each activity.

4. How many more people voted for hiking and biking combined than for riding a scooter?

$10 + 15 = 25$

$25 - 20 = 5$

5 more people voted for hiking and biking combined than for riding a scooter.

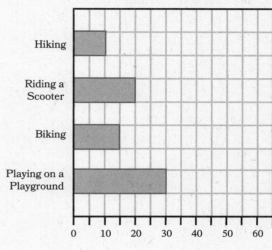

Favorite Outside Activity

Type of Activity

Number of Votes

I can use the bar graph to tell me how many students voted for each activity.

The right edge of each bar lines up with a number on the scale to show the number of votes.

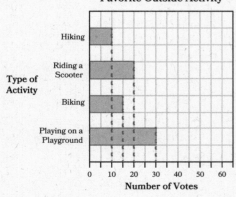

Favorite Outside Activity

First I need to find the total number of votes for hiking and biking.

I can add the number of votes for hiking, 10, and number of votes for biking, 15.

Then I subtract the number of votes for riding a scooter, 20, from the total.

Name _____

Use the array to help you fill in the blanks and find the totals.

1.

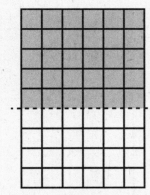

$5 \times 6 =$ _____

_____ $\times 6 =$ _____

$9 \times 6 = (5 +$ _____$) \times 6$

$= (5 \times 6) + ($ _____ $\times 6)$

$= 30 +$ _____

$=$ _____

Label the tape diagram. Then complete the equations.

2.

$(5 \times 8) =$ _____   _____ $= ($ _____ $\times 8)$

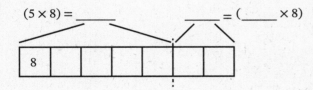

$7 \times 8 = (5 + 2) \times 8$

$= (5 \times 8) + ($ _____ $\times 8)$

$= 40 +$ _____

$=$ _____

3. Eva swims 7 laps every day for 6 days. How many total laps does she swim?

   a. To find the total, Casey breaks 6×7 into 5×7 and 1×7. Then she adds 35 and 7. Explain why her strategy works.

   b. Show another way 6×7 can be broken apart into smaller facts to find the product.

REMEMBER

Third grade students voted for their favorite inside activity.

The scaled bar graph shows how many students voted for each activity.

4. How many more people voted for reading and drawing combined than for doing a puzzle?

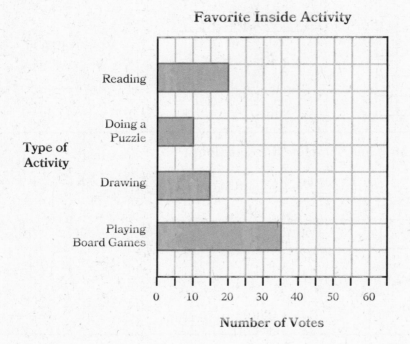

6

Name _____

Use the array to help you complete the number bond and equation.

1.

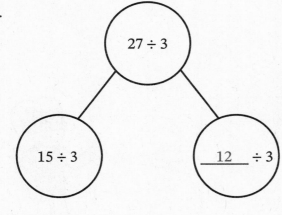

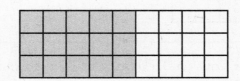

$$27 \div 3 = \underline{\quad 5 \quad} + \underline{\quad 4 \quad} = \underline{\quad 9 \quad}$$

15 $\underline{\quad 12 \quad}$

The number bond and array show how to break apart 27 into smaller parts, 15 and 12.

I can divide each part by 3.

$15 \div 3 = 5$ $\qquad$ $12 \div 3 = 4$

Then I add the quotients.

$5 + 4 = 9$

Use the array to help you complete the equation.

2.

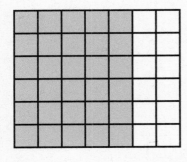

$42 ÷ 6 =$ ___5___ $+$ ___2___ $=$ ___7___

30   ___12___

The array shows how to break apart 42 into smaller parts, 30 and 12.

I can use familiar facts to help me divide each part by 6.

$30 ÷ 6 = 5$ $12 ÷ 6 = 2$

Then I add the quotients to find $42 ÷ 6$.

$5 + 2 = 7$

Use the break apart and distribute strategy to divide.

3. $72 ÷ 8 =$ ___5___ $+$ ___4___ $=$ ___9___

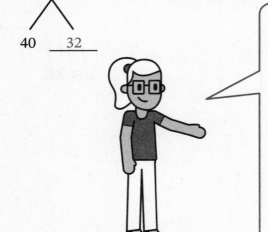

40   ___32___

To break apart 72, I think about facts I know well that can help me divide by 8.

I know $8 × 5 = 40$ and $8 × 4 = 32$.

I can use a number bond to help me break 72 into 40 and 32.

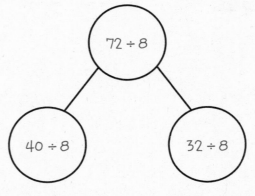

REMEMBER

Use the Read–Draw–Write process to solve the problem.

4. Robin buys a box of 40 baseball cards.

   The cards are equally divided into 5 packs.

   Robin gives 1 pack to his sister.

   How many baseball cards does he have left?

   $40 \div 5 = 8$

   $40 - 8 = 32$

   Robin has 32 baseball cards left.

I read the problem. I read again.

As I reread, I think about what I can draw.

I draw a tape diagram partitioned into 5 equal groups to represent the 5 packs of baseball cards. I label the total 40.

The size of each group is the unknown.

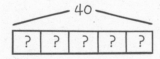

To help me find how many cards are in each pack, I can think: 5 groups of what number makes 40?

To find how many cards are left, I can subtract the number of cards in one pack, 8, from the number of cards Robin started with, 40.

6

Name _____

Use the array to help you complete the number bond and equation.

1.

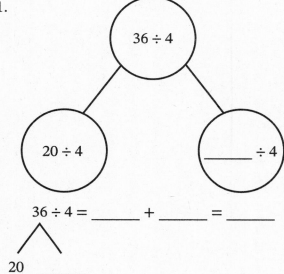

$36 \div 4 =$ _____ $+$ _____ $=$ _____

20 _____

Use the array to help you complete the equation.

2.

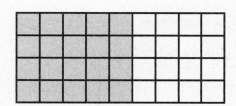

$48 \div 6 =$ _____ $+$ _____ $=$ _____

30 _____

Use the break apart and distribute strategy to divide.

3. $64 \div 8 =$ _____ $+$ _____ $=$ _____

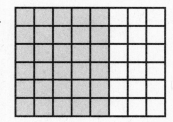

40 _____

4. $36 \div 6 =$ _____ $+$ _____ $=$ _____

30 _____

REMEMBER

Use the Read–Draw–Write process to solve the problem.

5. Luke buys a box of 20 chocolates.

   The box has 4 equal rows of chocolates.

   Luke eats 1 row of chocolates.

   How many chocolates are left in the box?

FAMILY MATH

Multiplication and Division Concepts with an Emphasis on the Unit of 7

Dear Family,

Your student is applying strategies for multiplying and dividing introduced earlier in the year, now with an emphasis on the unit 7. As your student learns their multiplication facts for 7, they use the break apart and distribute strategy to break the unit of 7 into more familiar parts. Your student also explores another strategy that is helpful when one factor is large. They write a multiplication expression with 2 factors as a multiplication expression with 3 factors. Then they use parentheses to group the 3 factors in different ways, and they learn that the order of multiplying the factors doesn't change the total.

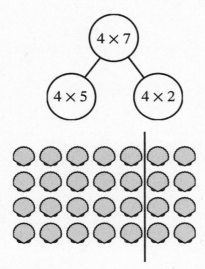

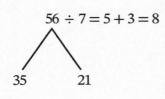

$$56 \div 7 = 5 + 3 = 8$$

35   21

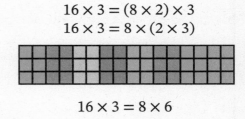

$$16 \times 3 = (8 \times 2) \times 3$$
$$16 \times 3 = 8 \times (2 \times 3)$$

$$16 \times 3 = 8 \times 6$$

The factors in a multiplication problem can be broken into smaller parts. Because $7 = 5 + 2$, fives and twos facts are helpful when multiplying by 7.

The total in a division problem can be broken into smaller parts. To divide by 7, students choose parts that can be divided by 7 with known facts.

The array shows 16×3. It also shows 8 groups of 3×2. Students can break apart arrays into smaller problems and use parentheses to group factors.

At-Home Activity

The Sevens Have It

Play a game with your student to help them practice multiplying by 7. Write the numbers 2–10 on small pieces of paper or index cards. Make one set for your student and one for yourself. Shuffle the numbers in each set. Place your student's numbers in a pile facedown in front of them and yours in a pile facedown in front of you. Ask your student to take the top paper from their pile and multiply the number by 7. Then you take the top paper from your pile and multiply the number by 7. Whoever has the larger product takes both numbers and puts them facedown at the bottom of their pile. For example,

- Your student picks a 5 from their pile. They say, "$7 \times 5 = 35$."

- You pick a 4 from your pile. You say "$7 \times 4 = 28$."

- Your student takes the papers with 5 and 4 and puts them facedown at the bottom of their pile because 35 is the larger product.

Repeat the process until one person has all the slips of paper in their pile.

Name _____

1. Use the array to skip-count by sevens. Then complete each equation.

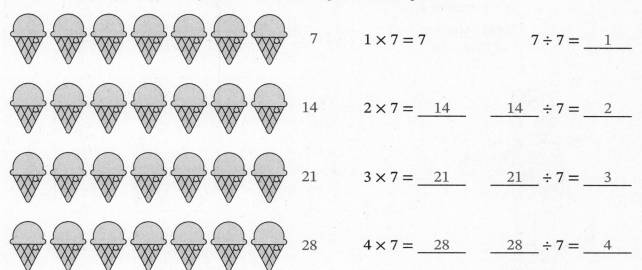

7        $1 \times 7 = 7$ $7 \div 7 = \underline{1}$

14       $2 \times 7 = \underline{14}$ $\underline{14} \div 7 = \underline{2}$

21       $3 \times 7 = \underline{21}$ $\underline{21} \div 7 = \underline{3}$

28       $4 \times 7 = \underline{28}$ $\underline{28} \div 7 = \underline{4}$

There are 7 ice cream cones in each row, and there are 4 rows.

I can skip-count by sevens 4 times to find the total number of ice cream cones.

2. Adam sorts crayons into containers. He puts 7 crayons in each container. The table shows how many crayons are needed for different numbers of containers. Complete the table.

Number of Containers	1	2	3	4	5	6
Total Number of Crayons	7	14	21	28	35	42

The number of containers increases by 1 each time.

I can skip-count by sevens to fill in the unknown numbers in the bottom row of the table.

Find the value of each unknown.

3. $63 \div y = 7$

   $y = \underline{\quad 9 \quad}$

4. $p \times 7 = 56$

   $p = \underline{\quad 8 \quad}$

5. $21 \div n = 7$

   $n = \underline{\quad 3 \quad}$

I can think about division problems as unknown factor problems. I know that $7 \times 9 = 63$. So, $63 \div 9 = 7$.

I can think about how many groups of 7 make 56 and skip-count by sevens to get to 56.

7, 14, 21, 28, 35, 42, 49, 56

I had to count by sevens 8 times to get to 56, so $8 \times 7 = 56$.

I know that $7 \times 3 = 21$. So, $21 \div 3 = 7$.

REMEMBER

6. How many liters of water are shown in the container?

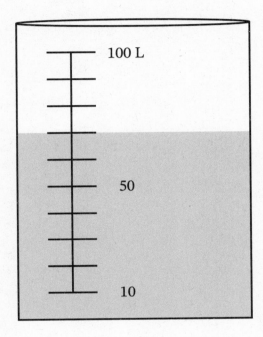

100 L

50

10

_____ 70 L _____

I see that the scale shows a liquid volume between 50 liters and 100 liters.

Each tick mark represents 10 liters.

The level of the water is 2 marks past 50 liters. I can count on from 50 by tens.

50, 60, 70

There are 70 L of water in the container.

Name

1. Use the array to skip-count by sevens. Then complete each equation.

$1 \times 7 = 7$ $7 \div 7 = \underline{\hspace{1cm}}$

$2 \times 7 = \underline{\hspace{1cm}}$ $14 \div 7 = \underline{\hspace{1cm}}$

$3 \times 7 = \underline{\hspace{1cm}}$ $\underline{\hspace{1cm}} \div 7 = \underline{\hspace{1cm}}$

$4 \times 7 = \underline{\hspace{1cm}}$ $\underline{\hspace{1cm}} \div 7 = \underline{\hspace{1cm}}$

$5 \times 7 = \underline{\hspace{1cm}}$ $\underline{\hspace{1cm}} \div 7 = \underline{\hspace{1cm}}$

2. Oka puts cookies in boxes. Each box has 7 cookies. The table shows how many cookies are needed for different numbers of boxes. Complete the table.

Number of Boxes	1	2	3	4		6	7	8	9
Total Number of Cookies	7			28				56	

Find the value of each unknown.

3. $7 \times d = 56$

    $d = \underline{\hspace{1cm}}$

4. $49 \div g = 7$

    $g = \underline{\hspace{1cm}}$

5. $m \times 7 = 42$

    $m = \underline{\hspace{1cm}}$

REMEMBER

6. How many liters of water are shown in the bottle?

_____

Name

Find 7×7 by using the break apart and distribute strategy. A line is drawn in the array to help you.

1.

$7 \times 7 = (7 \times 5) + (7 \times \underline{\quad 2 \quad})$

$7 \times 7 = 35 + (\underline{\quad 14 \quad})$

$7 \times 7 = \underline{\quad 49 \quad}$

$7 \times 5 = \underline{\quad 35 \quad}$ $7 \times 2 = \underline{\quad 14 \quad}$

There are 7 rows of 7 beach balls in the array.

The line breaks the array into 7 rows of 5 and 7 rows of 2.

I know that $7 \times 5 = 35$ and $7 \times 2 = 14$.

I can add 35 and 14 to find the total number of beach balls in the array. There are 49 beach balls in all.

Find 6 × 7 by using the break apart and distribute strategy. Complete the number bond to help you.

2.

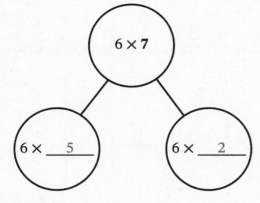

Sample:

$6 \times 7 = 6 \times (\underline{\quad 5 \quad} + \underline{\quad 2 \quad})$

$= (6 \times \underline{\quad 5 \quad}) + (6 \times \underline{\quad 2 \quad})$

$= \underline{\quad 30 \quad} + \underline{\quad 12 \quad}$

$= \underline{\quad 42 \quad}$

I can break apart the factor 7 into smaller, more familiar factors, 5 and 2.

Then I multiply to find each product.

I know 6 × 5 = 30 and 6 × 2 = 12.

I add 30 and 12 to find 6 × 7.

REMEMBER

3. Circle the number that correctly completes each statement.

An umbrella weighs about _____ grams.

20   (200)

A large dog weighs about _____ kilograms.

3   (30)

A cleaning bucket holds about _____ liters.

(10)   100

I can think about what 1 unit weighs to help me estimate.

• 1 gram is about the weight of a paper clip or a piece of gum.

• 1 kilogram is about the weight of a dictionary.

I can think about the amount of liquid in 1 liter to help me estimate.

• 1 liter is about the amount of water in a large sports water bottle.

• 4 liters is about the amount of liquid in a large jug of milk.

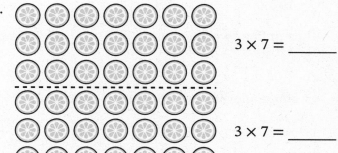

Name _____

Find 6×7 by using the break apart and distribute strategy in two different ways. A line is drawn in each array to help you.

1.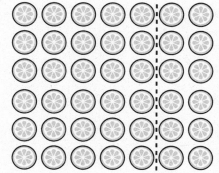

$6 \times 5 = $ _____ $6 \times 2 = $ _____

$6 \times 7 = (6 \times 5) + (6 \times $ _____ $)$

$6 \times 7 = 30 + $ _____

$6 \times 7 = $ _____

2.

$3 \times 7 = $ _____

$3 \times 7 = $ _____

$6 \times 7 = (3 \times 7) + ($ _____ $\times 7)$

$6 \times 7 = 21 + $ _____

$6 \times 7 = $ _____

Find 8×7 by using the break apart and distribute strategy in two different ways. Complete each number bond to help you.

3.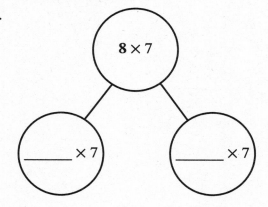

$8 \times 7 = ($ _____ $ + $ _____ $) \times 7$

$= ($ _____ $\times 7) + ($ _____ $\times 7)$

$= $ _____ $ + $ _____

$= $ _____

4.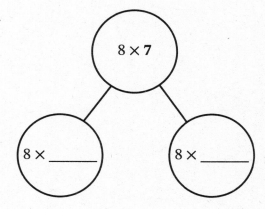

$8 \times 7 = 8 \times ($ _____ $ + $ _____ $)$

$= (8 \times $ _____ $) + (8 \times $ _____ $)$

$= $ _____ $ + $ _____

$= $ _____

REMEMBER

5. Circle the number that correctly completes each statement.

An apple weighs about _____ grams.

      19   190

A bag of flour weighs about _____ kilograms.

      2   20

A water cooler for a sports team holds about _____ liters.

      2   20

9

Name

Circle to show the equal groups in each array. Then circle the expression that represents the equal groups.

1. 6 groups of 2 × 4

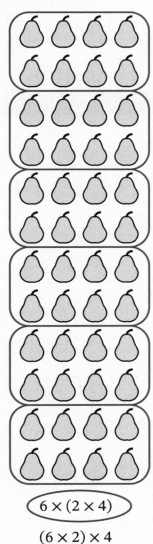

6 × (2 × 4)

(6 × 2) × 4

I circle 2 rows of 4 pears to show 1 group.

I draw 6 circles to show 6 groups.

There are 2 rows of 4 pears in each group.

The expression 6 × (2 × 4) represents my drawing.

I multiply 2 × 3 first. The product is 6.

Then I find 7 × 6.

Rewrite the expression with two factors. Then find the product.

2. 7 × (3 × 2)

7 × 6 = 42

Place parentheses in the equations to simplify and complete the problem.

3.　$3 \times 18 = 3 \times (2 \times 9)$

$= (3 \times 2) \times 9$

$= \underline{\ \ 6\ \ } \times \underline{\ \ 9\ \ }$

$= \underline{\ \ 54\ \ }$

> 18 is the product of 2 and 9, so I can rewrite 18 as 2 × 9 without changing the problem.
>
> I move the parentheses to smaller factors to make it easier for me to multiply.
>
> Now all I have to do is find 6 × 9.

REMEMBER

Use the Read–Draw–Write process to solve the problem.

4.　Oka makes 9 batches of potato salad. She uses 3 kilograms of potatoes in each batch.

How many kilograms of potatoes does she use in all?

$9 \times 3 = 27$

Oka uses 27 kilograms of potatoes in all.

> I read the problem. I read again.
>
> As I reread, I think about what I can draw.
>
> I draw a tape diagram with 9 equal parts to represent the 9 batches of potato salad.
>
> I know there are 3 kilograms of potatoes in each batch, so I label each part 3.
>
> The unknown is the total number of kilograms Oka uses.
>
>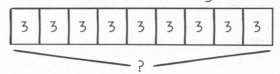
>
> I can multiply the number of groups, 9, by the number in each group, 3, to find the total number of kilograms of potatoes.

9

Name

Circle to show the equal groups in each array. Then circle the expression that represents the equal groups.

1. 4 groups of 3 × 4

$(4 \times 3) \times 4$

$4 \times (3 \times 4)$

2. 3 groups of 4 × 4

$3 \times (4 \times 4)$

$(4 \times 3) \times 4$

Rewrite the expression with two factors. Then find the product.

3. $6 \times (3 \times 3)$

4. $(2 \times 4) \times 2$

Place parentheses in the equations to simplify and complete the problem.

5. $4 \times 14 = 4 \times (2 \times 7)$

$= 4 \times 2 \times 7$

$= \underline{\hspace{1cm}} \times \underline{\hspace{1cm}}$

$= \underline{\hspace{1cm}}$

REMEMBER

Use the Read–Draw–Write process to solve the problem.

6. Ivan makes 8 batches of salsa. He uses 4 kilograms of tomatoes in each batch.

How many kilograms of tomatoes does he use in all?

10

_____
Name

Complete each equation.

1. $15 - (8 + 2) =$ ___5___

    $(15 - 8) + 2 =$ ___9___

2. $(7 + 9) - 2 =$ ___14___

    $7 + (9 - 2) =$ ___14___

> I complete the problem inside the parentheses first. Sometimes, moving the parentheses changes the value of an expression.
>
> $$15 - (8 + 2) = 15 - 10 = 5$$
> $$(15 - 8) + 2 = 7 + 2 = 9$$
>
> The values are not equal because 5 does not equal 9.
>
> Other times, moving the parentheses does not change the value of an expression.
>
> $$(7 + 9) - 2 = 16 - 2 = 14$$
> $$7 + (9 - 2) = 7 + 7 = 14$$
>
> The values are equal because they are both 14.

3. Create two different expressions by grouping different parts of $6 \times 6 + 3$ by using parentheses. Then find their values.

    $(6 \times 6) + 3 =$ ___39___

    $6 \times (6 + 3) =$ ___54___

> I can put the parentheses around 6×6, or I can put them around $6 + 3$.
>
> For $(6 \times 6) + 3$, I find 6×6 first and then add 3.
>
> For $6 \times (6 + 3)$, I find $6 + 3$ first and then multiply by 6.
>
> For this problem, moving the parentheses changes the value of the expression.

Use parentheses to make the equation true.

4. $8 = 8 \times (5 - 4)$

For 8 × 5 - 4, first I try placing the parentheses around 8 × 5.

That means I multiply first and then subtract.

$$(8 \times 5) - 4 = 40 - 4 = 36$$

I am looking for a value of 8, so this is not correct.

Next I try placing the parentheses around 5 - 4.

I subtract first and then multiply.

$$8 \times (5 - 4) = 8 \times 1 = 8$$

REMEMBER

Find the unknown.

5. $443 - 156 = \underline{\quad 287 \quad}$

I can use a place value chart to subtract. I check the total to see if I am ready to subtract.

I unbundle 1 ten into 10 ones.

I unbundle 1 hundred into 10 tens.

Now I am ready to subtract.

100s	10s	1s
• • • •	• • • •	• • •

100s	10s	1s
• • • ⤫	• • • ⤫	• • •
	• • • • •	• • • • •
	• • • • •	• • • • •

100s	10s	1s
• • ⤫	⤫⤫⤫⤫	⤫⤫⤫
	• • • • •	• • • • •
	• • • • •	• • • • •

6. $413 + 398 = \underline{\quad 811 \quad}$

I see that 398 is close to a benchmark number, 400.

I can use the make the next hundred strategy to find the sum.

I decompose 413 into 2 and 411 and add 2 to 398 to make 400.

$$413 + 398 = 411 + 400 = 811$$

411 2

Now I have a simpler addition problem.

10

Name _____

Complete each equation.

1. $(12 - 3) + 7 =$ _____

   $12 - (3 + 7) =$ _____

2. $(8 \div 2) + 2 =$ _____

   $8 \div (2 + 2) =$ _____

3. $5 \times (4 + 1) =$ _____

   $(5 \times 4) + 1 =$ _____

4. $12 + (2 \times 2) =$ _____

   $(12 + 2) \times 2 =$ _____

Complete each equation. Circle the pairs that have the same value for both equations.

5. $9 + (5 + 2) =$ _____

   $(9 + 5) + 2 =$ _____

6. $2 \times (4 \times 2) =$ _____

   $(2 \times 4) \times 2 =$ _____

7. $(20 \div 2) + 3 =$ _____

   $20 \div (2 + 3) =$ _____

8. $(14 - 6) + 4 =$ _____

   $14 - (6 + 4) =$ _____

9. Create two different expressions by grouping different parts of $5 \times 6 + 4$ by using parentheses. Then find their values.

   $5 \times 6 + 4 =$ _____

   $5 \times 6 + 4 =$ _____

Use parentheses to make each equation true.

10. $32 = 7 \times 5 - 3$

11. $8 \div 2 \times 6 = 24$

12. $17 + 8 \times 2 = 33$

13. $28 - 10 + 5 = 23$

14. David says the value of $3 \times 6 \div 2$ is 9 no matter where he puts the parentheses. Is he correct? Place parentheses around different numbers to explain his thinking.

Find the unknown.

15. $857 - 269 =$ _____

16. $629 + 178 =$ _____

11

Name _____

1. Complete the equations in the number bond to find 54 ÷ 6. Use each part of the array to help divide.

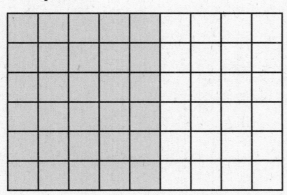

$30 \div 6 = \underline{5}$

$24 \div 6 = \underline{4}$

$54 \div 6 = \underline{9}$

> There are 54 squares in the array. I can use the break apart and distribute strategy to find 54 ÷ 6.
>
> I can decompose the total into smaller parts. I know 5 sixes is 30 and 4 sixes is 24.
>
> 5 and 4 make 9. So, 54 ÷ 6 = 9.

Divide by using the break apart and distribute strategy.

$49 \div 7 = \underline{5} + \underline{2} = \underline{7}$

35   14

> I can use the break apart and distribute strategy to find 49 ÷ 7.
>
> I know my fives facts well, so I decompose 49 into 35 and 14.
>
> I know 35 ÷ 7 is 5 and 14 ÷ 7 is 2.
>
> I add the quotients to find 49 ÷ 7.

$(35 \div 7) + (14 \div 7)$

$49 \div 7 = \underline{5} + \underline{2} = \underline{7}$

35   14

Use the Read–Draw–Write process to solve the problem.

2. Mia has 72 playing cards. She puts them in rows of 8.

   How many rows of cards does Mia have?

   $72 \div 8 = 9$

   Mia has 9 rows of playing cards.

> I read the problem. I read again.
>
> As I reread, I think about what I can draw.
>
> I draw a tape diagram to represent the playing cards and label the total. There are 8 cards in each row, so I know the size of each group.
>
>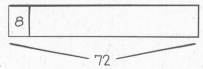
>
> To help me solve, I think about how many eights are in 72.
>
> I draw a number bond to decompose 72 into smaller parts. I break it into a fives fact and another fact.
>
>
>
> I know 40 ÷ 8 = 5 and 32 ÷ 8 = 4. I add the quotients to find 72 ÷ 8.

REMEMBER

Add. Label each addend and the sum with even (e) or odd (o).

3. $13 + 4 = \underline{\quad 17 \quad}$
       o   e   o

I say even numbers when I skip-count by twos.

2, 4, 6, 8, 10, 12, 14, 16, 18, 20

I do not say 13 when I skip-count by twos, so 13 is odd.

I say 4 when I skip-count by twos, so 4 is even.

I do not say 17 when I skip-count by twos, so 17 is odd.

An odd number plus an even number equals an odd number.

11

Name _____

1. Complete the equations in the number bond to find $63 \div 7$. Use each part of the array to help divide.

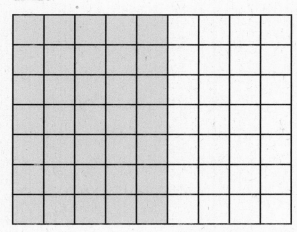

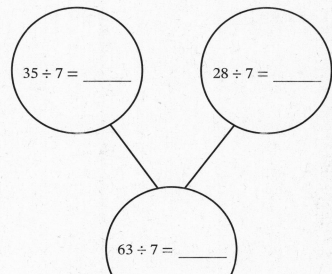

$35 \div 7 =$ _____

$28 \div 7 =$ _____

$63 \div 7 =$ _____

Divide by using the break apart and distribute strategy.

2. $48 \div 8 =$ _____ + _____ = _____

_____   _____

Use the Read–Draw–Write process to solve the problem.

3. Ivan plants 42 flowers. He plants them in rows of 6.

How many rows of flowers does Ivan have?

REMEMBER

Add. Label each addend and the sum with even (e) or odd (o).

4. $6 + 9 =$ _____

12

Name _____

Use the Read–Draw–Write process to solve each problem. Use a letter to represent the unknown.

1. There are 4 rows of chairs. Each row has 7 chairs.

   How many chairs are there in all?

   $4 \times 7 = m$

   $28 = m$

   There are 28 chairs in all.

   I read the problem. I read again.

   As I reread, I think about what I can draw.

   I draw an array with 4 rows of 7. I can skip-count by sevens to find the total number of chairs.

   7
   14
   21
   28

2. At the store, a display of soup cans has 6 equal rows. There are 72 soup cans in all.

   How many cans are in each row?

   $72 \div 6 = c$

   $12 = c$

   There are 12 cans of soup in each row.

   I read the problem. I read again.

   As I reread, I think about what I can draw.

   I draw a tape diagram to represent the 72 soup cans. I make 6 equal parts in the tape diagram to represent the 6 equal rows of cans.

   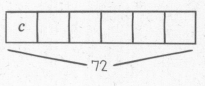

   The unknown is the size of each part. To help me solve, I ask myself: 6 groups of what number is 72?

   I draw a number bond to break 72 into smaller parts. I can break it into a tens fact and another fact.

   I break 72 into 60 and 12 because I know $60 \div 6 = 10$ and $12 \div 6 = 2$.

   72

   60   12

REMEMBER

Use the Read–Draw–Write process to solve the problem.

3. There are 3 groups of bananas.

    There are 5 bananas in each group.

    How many bananas are there in all?

    $3 \times 5 = 15$

    There are 15 bananas in all.

I read the problem. I read again.

As I reread, I think about what I can draw.

I draw 3 circles to show the groups of bananas.

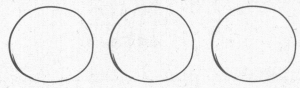

I draw 5 dots in each circle to show the bananas in each group.

12

Name _____

Use the Read–Draw–Write process to solve each problem. Use a letter to represent the unknown.

1. There are 6 tables in the library. Each table has 8 chairs. How many chairs are there in all?

2. A bookcase has 3 shelves. There are 9 books on each shelf. How many books are there in all?

3. Coach Diaz wants to put basketballs equally into bins. There are 32 basketballs in all. She has 4 bins. How many basketballs go in each bin?

REMEMBER

Use the Read–Draw–Write process to solve the problem.

4. There are 8 bags of apples.

   There are 5 apples in each bag.

   How many apples are there in all?

FAMILY MATH

Analysis of Patterns Using Units of 9, 0, and 1

Dear Family,

Your student is exploring patterns when multiplying and dividing with units of 9, 0, and 1. They use the relationship between 9 and 10 to help them skip-count and multiply by 9. They also identify special patterns that help them better remember nines facts. They investigate multiplication and division patterns with 1 and 0 and discover that there are special rules when multiplying and dividing with these units. Using their knowledge of multiplication and division, they analyze patterns in multiplication tables and use patterns to find unknown numbers. They also write their own word problems and solve word problems with two steps.

Key Term

multiple

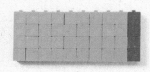

9 fours = 10 fours − 1 four

$9 \times 4 = (10 \times 4) - (1 \times 4)$

$= 40 - 4$

$= 36$

1 nine is 9
2 nines is 18
3 nines is 27
4 nines is 36
5 nines is 45
6 nines is 54
7 nines is 63
8 nines is 72
9 nines is 81
10 nines is 90

$\boxed{7}$ groups of 0 is ___0___ .

$\boxed{7} \times$ ___0___ $=$ ___0___

0 groups of $\boxed{7}$ is ___0___ .

___0___ $\times \boxed{7} =$ ___0___

$1 \times 2 = 2$
$1 \times 3 = 3$
$1 \times 4 = 4$
$1 \times 5 = 5$
$1 \times 6 = 6$
$1 \times 7 = 7$
$1 \times 8 = 8$
$1 \times 9 = 9$

$1 \times 30 = 30$
$1 \times 500 = 500$

To find 9 units of four, find 10 units of four and then subtract 1 unit of four.

The number of tens in a multiple of 9 is 1 less than the number of groups of 9.

Any number multiplied by 0 is 0. A number cannot be divided by 0, but 0 divided by any other number is 0.

Any number, except 0, multiplied or divided by 1 is itself.

At-Home Activity

One Group Less

Encourage your student to practice multiplying with units of 9 by using the strategy of multiplying by 10 and subtracting 1 group. Look for an object with a set number of features, such as a fork with 4 prongs or a window with 6 panes. Show it to your student and ask them questions to help them find the total number of features for 9 of those objects.

- "If 1 fork has 4 prongs, how many prongs are on 10 forks?" ($10 \times 4 = 40$)

- "If there are 40 prongs on 10 forks, how many prongs are on 9 forks?" ($40 - 4 = 36$)

If possible, place 10 of the object in front of your student and then remove 1 when you ask about 9 objects.

13

Name ______________________________

1. Fill in the blanks to help you skip-count by nines.

0	$\xrightarrow{+10}$	10	$\xrightarrow{-1}$	(9)
9	$\xrightarrow{+10}$	19	$\xrightarrow{-1}$	(18)
18	$\xrightarrow{+10}$	28	$\xrightarrow{-1}$	(27)
27	$\xrightarrow{+10}$	37	$\xrightarrow{-1}$	(36)
36	$\xrightarrow{+10}$	46	$\xrightarrow{-1}$	(45)
45	$\xrightarrow{+10}$	55	$\xrightarrow{-1}$	(54)
54	$\xrightarrow{+10}$	64	$\xrightarrow{-1}$	(63)
63	$\xrightarrow{+10}$	73	$\xrightarrow{-1}$	(72)
72	$\xrightarrow{+10}$	82	$\xrightarrow{-1}$	(81)
81	$\xrightarrow{+10}$	91	$\xrightarrow{-1}$	(90)

I can add 10 and subtract 1 to skip-count by nines.

For example, $3 \times 9 = 27$.

10 more than 27 is 37. 1 less than 37 is 36.

This helps me name all the multiples of 9.

The **multiples** of 9 are the numbers we say when we skip-count by nines.

Multiples of 9 are also the products we find when we multiply 9 by other numbers.

For example, 18 is a multiple of 9 because $18 = 2 \times 9$.

2. David writes $6 \times 9 = 54$.

a. He checks his work by thinking about a sum of 5 and 4. Explain David's strategy.

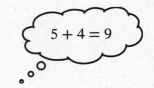

5 + 4 = 9

David thinks about the sum of 5 and 4 to check his work because he knows that when you multiply by 9, the sum of the digits in the product equals 9.

b. Did David correctly multiply 6×9? How do you know?

David multiplied correctly. I know because $5 \times 9 = 45$. I added 10 to 45 and then subtracted 1 to find 6×9. So, $6 \times 9 = 54$.

I can use multiplication facts I know to help me with facts that I do not know.

I think about 6×9 as 6 nines.

I know 5 nines is 45.

Then I add 1 more nine.

I can add 9 efficiently by adding 10 and subtracting 1.

---

3. Find the product. Describe the strategy that you used.

$4 \times 9 = 36$

I know that $3 \times 9 = 27$, so I added 10 to 27 and subtracted 1 to get 36.

I can use a product that I know to help find the answer to 4×9.

I know that $3 \times 9 = 27$.

I add 10 to 27 to get 37, and subtract 1 to get 36.

I can find the product another way.

I can use a tens fact to help me find the product. I know that 4 tens is 40. Then all I have to do is subtract 4 from 40 to find 4×9.

REMEMBER

Round the number to the nearest ten and nearest hundred.

4.

	Rounded to the Nearest Ten	Rounded to the Nearest Hundred
848	850	800

To round to the nearest ten, I think about how many tens are in the number and what the next ten is. 848 has 84 tens. The next ten is 85 tens. So, 848 is between 840 and 850.

I can draw a vertical number line to show my thinking.

Then I think about what number is halfway between the tens. 845 is halfway between 840 and 850.

I ask myself which ten my number is closer to.

848 is more than halfway to the next ten, so 848 is closer to 850.

850 = 85 tens
848 = 84 tens 8 ones
845 = 84 tens 5 ones
840 = 84 tens

To round to the nearest hundred, I think about how many hundreds are in the number and what the next hundred is. 848 has 8 hundreds. The next hundred is 9 hundreds.

I can draw a vertical number line to show my thinking.

Then I think about what number is halfway between the hundreds.

850 is halfway between 800 and 900.

I ask myself which hundred my number is closer to.

848 is less than halfway to the next hundred, so 848 is closer to 800.

900 = 9 hundreds
850 = 8 hundreds 5 tens
848 = 8 hundreds 4 tens 8 ones
800 = 8 hundreds

13

Name _____

1. Fill in the blanks to help you skip-count by nines.

_____ $\xrightarrow{+10}$ 10 $\xrightarrow{-1}$ ⑨

9 $\xrightarrow{+10}$ 19 $\xrightarrow{-1}$ ◯

_____ $\xrightarrow{+10}$ _____ $\xrightarrow{-1}$ ◯

_____ $\xrightarrow{+10}$ 37 $\xrightarrow{-1}$ ◯

_____ $\xrightarrow{+10}$ _____ $\xrightarrow{-1}$ ◯

_____ $\xrightarrow{+10}$ _____ $\xrightarrow{-1}$ ◯

_____ $\xrightarrow{+10}$ _____ $\xrightarrow{-1}$ ◯

_____ $\xrightarrow{+10}$ _____ $\xrightarrow{-1}$ ⑦②

72 $\xrightarrow{+10}$ _____ $\xrightarrow{-1}$ ◯

_____ $\xrightarrow{+10}$ 91 $\xrightarrow{-1}$ ◯

2. Robin writes $7 \times 9 = 62$.

   a. She checks her work by thinking about the sum of 6 and 2. Explain Robin's strategy.

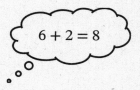

   $6 + 2 = 8$

   b. Did Robin correctly multiply 7 and 9? How do you know?

3. Find the product. Describe the strategy that you used.

   $8 \times 9 =$ _____

REMEMBER

4. Round the number to the nearest ten and nearest hundred.

	Rounded to the Nearest Ten	Rounded to the Nearest Hundred
183		
415		
597		

14

Name

1. Draw lines to match. Fill in the blanks to complete the expressions.

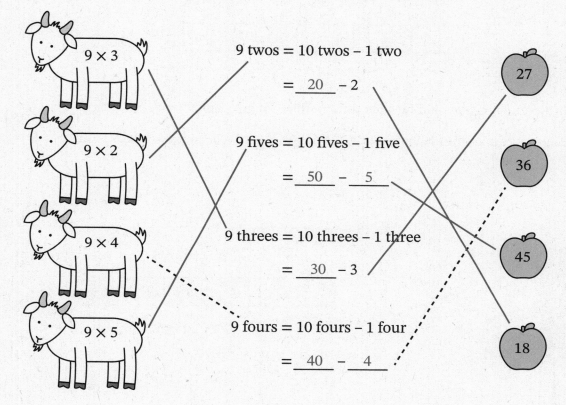

9 × 3

9 × 2

9 × 4

9 × 5

9 twos = 10 twos − 1 two

     = __20__ − 2

9 fives = 10 fives − 1 five

     = __50__ − __5__

9 threes = 10 threes − 1 three

     = __30__ − 3

9 fours = 10 fours − 1 four

     = __40__ − __4__

27

36

45

18

To find the product of 9 and another number, I can use a tape diagram and the 9 = 10 − 1 strategy.

To find 9 × 5, I first make a tape diagram with 10 fives.

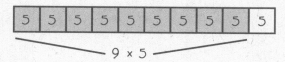

5   5   5   5   5   5   5   5   5   5

⎣____ 9 × 5 ____⎦

I know that 10 fives is 50, or 10 × 5 = 50.

I then subtract 1 five to get 45, or 50 − 5 = 45.

So, 9 × 5 = 45.

2. Adam has a box of golf balls. The box has 9 rows with 2 golf balls in each row. He uses 10 twos to find the total number of golf balls.

   a. Draw a model to represent Adam's strategy.

   Sample:

   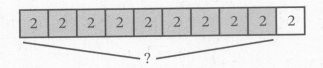

   b. Explain Adam's strategy and find the total number of golf balls.

   Adam's strategy is to find 10 twos and then subtract 1 two because 10 twos − 1 two = 9 twos. $20 − 2 = 18$, so $9 \times 2 = 18$.

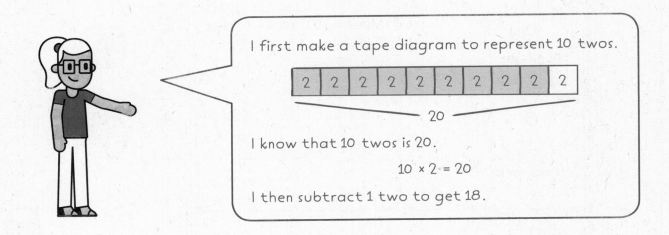

REMEMBER

Add. Show how you know.

3. 590 + 195 = __785__

Sample:

590 $\xrightarrow{+200}$ 790 $\xrightarrow{-5}$ 785

I notice that 195 is close to the benchmark number 200. I start at 590. I add 200 to get 790.

I can show my thinking on an open number line.

Since I added 200 instead of 195, I need to subtract 5.

I subtract 5 from 790 and get 785.

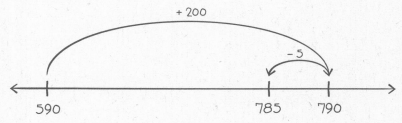

I can also show this strategy using the arrow way.

590 $\xrightarrow{+200}$ 790 $\xrightarrow{-5}$ 785

_/ **14**

Name

1. Draw lines to match. Fill in the blanks to complete the expressions.

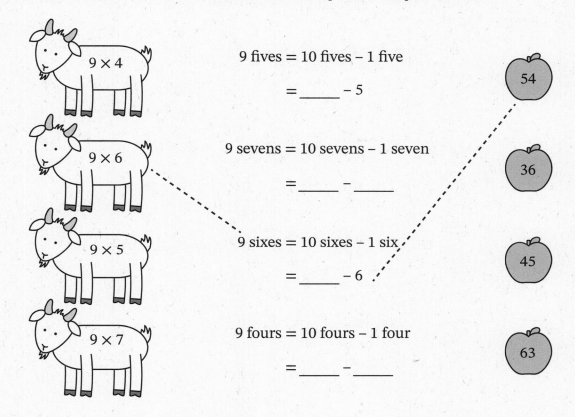

9 fives = 10 fives − 1 five

= _____ − 5

9 sevens = 10 sevens − 1 seven

= _____ − _____

9 sixes = 10 sixes − 1 six

= _____ − 6

9 fours = 10 fours − 1 four

= _____ − _____

2. Liz buys a box of markers. The box has 9 rows with 3 markers in each row. She uses 10 threes to find the total number of markers in the box.

a. Draw a model to represent Liz's strategy.

b. Explain Liz's strategy and find the total number of markers.

REMEMBER

Add. Show how you know.

3. $267 + 390 =$ _____

15

Name _____

Use the equal groups pictures to fill in the blanks.

1. 5 groups of 1 is ___5___ .      5 divided into groups of 1 is ___5___ groups.

    $5 \times 1 =$ ___5___ $5 \div 1 =$ ___5___

2. 7 groups of 1 is ___7___ .      7 divided into 7 equal groups is ___1___ in each group.

    $7 \times 1 =$ ___7___ $7 \div 7 =$ ___1___

3. 1 group of 4 is ___4___ .   4 divided into 1 group is ___4___ in each group.

    $1 \times 4 =$ ___4___ $4 \div 1 =$ ___4___

> I see 5 groups with 1 bird in each group. I count a total of 5 birds.
>
> I know that a number times 1 is itself.
>
> 5 × 1 = 5
>
> I see 7 groups with 1 butterfly in each group. I count a total of 7 butterflies.
>
> I know that a number divided by itself is 1. This rule does not work for 0, though.
>
> 7 ÷ 7 = 1
>
> I see 1 group of 4 fish. I count a total of 4 fish.
>
> I know that a number divided by 1 is itself.
>
> 4 ÷ 1 = 4

4.　Write whether each equation is true or false.

Equation	True or False
$6 \times 0 = 6$	False
$1 \times 8 = 8$	True
$0 \div 9 = 0$	True
$3 \div 3 = 0$	False

I know that any number multiplied by 0 is 0. $6 \times 0 = 6$ is false. The correct equation is $6 \times 0 = 0$.

I know that any number times 1 is itself. The equation $1 \times 8 = 8$ is true.

I know that 0 divided by any number except for 0 is 0. The equation $0 \div 9 = 0$ is true.

I know that any number divided by itself is 1. The equation $3 \div 3 = 0$ is false. The correct equation is $3 \div 3 = 1$.

5.　Explain why the equation $7 \times 1 = 1$ is false.

$7 \times 1 = 1$ is false because any number times 1 is itself. It should be $7 \times 1 = 7$.

I know that any number times 1 is itself.

The equation $7 \times 1 = 1$ is false.

The correct equation is $7 \times 1 = 7$.

REMEMBER

Find the unknown.

6. 475 − 240 = __235__

I know 475 is 4 hundreds 7 tens 5 ones and 240 is 2 hundreds 4 tens.

I subtract the hundreds.

4 hundreds – 2 hundreds = 2 hundreds

I subtract the tens.

7 tens – 4 tens = 3 tens

240 doesn't have any ones to subtract, so the ones stay the same.

The answer is 2 hundreds 3 tens 5 ones, or 235.

So, 475 – 240 = 235.

Name _____

Use the equal groups pictures to fill in the blanks.

1. 3 groups of 1 is _____ .

    $3 \times 1 =$ _____

    3 divided into groups of 1 is _____ groups.

    $3 \div 1 =$ _____

2. 8 groups of 1 is _____ .

    $8 \times 1 =$ _____

    8 divided into 8 equal groups is _____ in each group.

    $8 \div 8 =$ _____

3. 1 group of 6 is _____ .

    $1 \times 6 =$ _____

    6 divided into 1 group is _____ in each group.

    $6 \div 1 =$ _____

4. Write whether each equation is true or false.

Equation	True or False
$7 \times 0 = 0$	
$1 \times 4 = 1$	
$6 \div 6 = 1$	
$5 \div 0 = 0$	

5. Choose one of the false statements from problem 4 and explain why it is false.

REMEMBER

Find the unknown.

6. $883 - 320 =$ _____

Name

Use your completed multiplication table to answer problems 1–4.

1. Decide whether each pattern is true or false. Write an equation that supports your decision.

Pattern	True or False	Equation
odd times odd equals even	False	Sample: $3 \times 7 = 21$
even times even equals odd	False	Sample: $6 \times 8 = 48$
even times odd equals even	True	Sample: $4 \times 7 = 28$

For "even times odd equals even," I look for where an even-numbered row meets an odd-numbered column.

The row with 4 and the column with 7 meet at 28, which is even.

So "even times odd equals even" is true.

×	1	2	3	4	5	6	7	8	9	10
1	1	2	3	4	5	6	7	8	9	10
2	2	4	6	8	10	12	14	16	18	20
3	3	6	9	12	15	18	21	24	27	30
4	4	8	12	16	20	24	28	32	36	40
5	5	10	15	20	25	30	35	40	45	50
6	6	12	18	24	30	36	42	48	54	60
7	7	14	21	28	35	42	49	56	63	70
8	8	16	24	32	40	48	56	64	72	80
9	9	18	27	36	45	54	63	72	81	90
10	10	20	30	40	50	60	70	80	90	100

For "odd times odd equals even," I look for where an odd-numbered row meets an odd-numbered column.

The row with 3 and the column with 7 meet at 21, which is odd.

So "odd times odd equals even" is false.

For "even times even equals odd," I look for where an even-numbered row meets an even-numbered column.

The row with 6 and the column with 8 meet at 48, which is even.

So "even times even equals odd" is false.

2. Circle the product 72 in the multiplication table. Explain why the product 72 is in the table more than once.

The product 72 is in the table more than once because 8 × 9 = 72 and 9 × 8 = 72.

Once I find 72, I can look at the labels for the row and column.

I see 9 is the label of the row and 8 is the label of the column. So 9 × 8 = 72.

I know I can switch the order of the factors to get a second fact with the same product. So 8 × 9 = 72.

×	1	2	3	4	5	6	7	8	9	10
1	1	2	3	4	5	6	7	8	9	10
2	2	4	6	8	10	12	14	16	18	20
3	3	6	9	12	15	18	21	24	27	30
4	4	8	12	16	20	24	28	32	36	40
5	5	10	15	20	25	30	35	40	45	50
6	6	12	18	24	30	36	42	48	54	60
7	7	14	21	28	35	42	49	56	63	70
8	8	16	24	32	40	48	56	64	(72)	80
9	9	18	27	36	45	54	63	(72)	81	90
10	10	20	30	40	50	60	70	80	90	100

3. Pablo says, "I can use the multiplication table to find 6 × 18."

Deepa says, "But 6 × 18 is not in the table."

Explain how Pablo might use the table to find 6 × 18.

Sample:

Pablo might use the table to find 6 × 18 by breaking the expression apart into facts that are in the table. He could break it apart into 6 × 10 and 6 × 8. Then he could find those products and add them together to find 6 × 18.

I can find 6 × 18 by breaking the expression apart into smaller facts that are in the table.

I can break 6 × 18 apart into 6 × 10 and 6 × 8.

I can find those products and add them together to find 6 × 18.

60 + 48 = 108

4. How can you use the table to find 63 ÷ 7?

Sample:

You can use the table to find 63 ÷ 7 by sliding your finger across the 7 row until you get to 63. When you get to 63, you see what number is at the top of that column, and that number is the answer.

To find 63 ÷ 7, I slide my finger across the 7 row to 63.

When I get to 63, I slide up the column to the top. The number there is the quotient.

The number at the top of the column is 9.

So 63 ÷ 7 = 9.

×	1	2	3	4	5	6	7	8	9	10
1	1	2	3	4	5	6	7	8	9	10
2	2	4	6	8	10	12	14	16	18	20
3	3	6	9	12	15	18	21	24	27	30
4	4	8	12	16	20	24	28	32	36	40
5	5	10	15	20	25	30	35	40	45	50
6	6	12	18	24	30	36	42	48	54	60
7	7	14	21	28	35	42	49	56	63	70
8	8	16	24	32	40	48	56	64	72	80
9	9	18	27	36	45	54	63	72	81	90
10	10	20	30	40	50	60	70	80	90	100

REMEMBER

5. Fill in the blanks to make the equations true.

$42 ÷ \underline{\quad 7 \quad} = 6$

$\underline{\quad 3 \quad} × 8 = 24$

To find $42 ÷ \underline{\quad\quad} = 6$, I can think about how many sixes are in 42.

To find $\underline{\quad\quad} × 8 = 24$, I can think about how many eights are in 24.

16

Name _____

Use the completed multiplication table to answer problems 1–4.

1. Decide whether each pattern is true or false. Write an equation that supports your decision.

×	1	2	3	4	5	6	7	8	9	10
1	1	2	3	4	5	6	7	8	9	10
2	2	4	6	8	10	12	14	16	18	20
3	3	6	9	12	15	18	21	24	27	30
4	4	8	12	16	20	24	28	32	36	40
5	5	10	15	20	25	30	35	40	45	50
6	6	12	18	24	30	36	42	48	54	60
7	7	14	21	28	35	42	49	56	63	70
8	8	16	24	32	40	48	56	64	72	80
9	9	18	27	36	45	54	63	72	81	90
10	10	20	30	40	50	60	70	80	90	100

Pattern	True or False	Equation
even times even equals even		
even times odd equals odd		
odd times even equals even		

2. Circle the product 45 in the multiplication table. Explain why the product 45 is in the table more than once.

3. Casey says, "I can use the multiplication table to find 8×15."

   James says, "But 8×15 is not in the table."

   Explain how Casey might use the table to find 8×15.

4. How can you use the table to find $54 \div 9$?

REMEMBER

5. Fill in the blanks to make the equations true.

$28 \div \underline{\hspace{1cm}} = 7$

$\underline{\hspace{1cm}} \times 9 = 27$

17

Name

1. A pentagon has 5 sides. Complete the table.

Number of Pentagons	1	2	3	4	5	6	7
Total Number of Sides	5	10	15	20	25	30	35

> 1 pentagon has 5 sides, and 2 pentagons have 10 sides.
>
> The number in the top row is multiplied by 5 to get the number in the bottom row.
>
> I notice a pattern in the bottom row. It shows skip-counting by fives.

2. Mr. Davis puts oranges in crates. Each crate has the same number of oranges.

   a. Complete the table.

Number of Oranges	9	18	27	36	45	54	63
Total Number of Crates	1	2	3	4	5	6	7

   b. How many crates does Mr. Davis need if he has 72 oranges?

   Mr. Davis needs 8 crates if he has 72 oranges. Each crate has 9 oranges, and $8 \times 9 = 72$.

> There are 9 oranges in 1 crate and 18 oranges in 2 crates.
>
> The number in the bottom row can be multiplied by 9 to find the number in the top row.

Write the pattern and complete the table.

3. Pattern: _Divide the input by 7_

Input	Output
35	5
14	2
42	6
49	7
7	1
21	3

I look for a pattern between the numbers in the input column and the numbers in the output column.

I see that when 7 is the input number, 1 is the output number. I see that when 14 is the input number, 2 is the output number.

I see that when 35 is the input number, 5 is the output number.

That means each input number can be divided by 7 to get the output number.

I divide the rest of the input numbers by 7 to find the output numbers.

REMEMBER

Use the Read–Draw–Write process to solve the problem.

4. James uses 58 blocks to build a tower.

   James uses 19 fewer blocks than Mia.

   How many blocks does Mia use?

   $58 + 19 = 77$

   Mia uses 77 blocks.

I read the problem. I read again.

As I reread, I think about what I can draw.

I can draw a tape diagram.

I draw a tape to represent James's blocks and label it 58.

J | 58 |

I know James uses 19 fewer blocks than Mia. So that means Mia uses 19 more blocks than James.

I draw a longer tape to represent Mia's blocks and label it with a question mark to show that it is unknown.

I know the difference in the number of blocks is 19, so I label the space between James's tape and Mia's tape 19.

The tape diagram helps me see that I can add 58 and 19 to find Mia's blocks. I can use the arrow way to show my thinking.

$$58 \xrightarrow{+20} 78 \xrightarrow{-1} 77$$

Name _____

1. A square has 4 sides. Complete the table.

Number of Squares	1	2	3	4	5	6	7
Total Number of Sides	4	8				24	

2. Miss Wong puts apples in boxes. Each box has the same number of apples.

a. Complete the table.

Number of Apples	8	16			40		
Total Number of Boxes	1	2	3	4	5	6	7

b. How many boxes does Miss Wong need if she has 72 apples?

Write the pattern and complete the table.

3. Pattern: _____

Input	Output
24	
8	2
16	4
28	
4	1
12	3

REMEMBER

Use the Read–Draw–Write process to solve the problem.

4. Casey picks 64 strawberries.

   Casey picks 17 fewer strawberries than Ivan.

   How many strawberries does Ivan pick?

18

Name _____

1. Write a word problem that can be represented with the expression 6 × 10. Use the picture to help you.

Sample: There are 6 cups of pencils. Each cup has 10 pencils. How many pencils are there in all?

_____

_____

_____

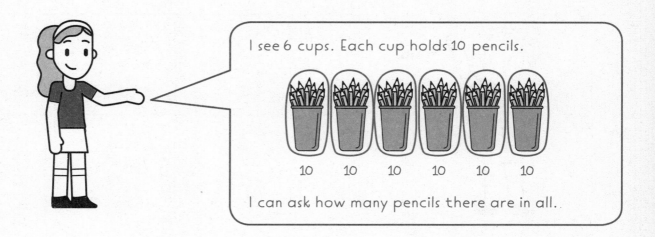

I see 6 cups. Each cup holds 10 pencils.

10   10   10   10   10   10

I can ask how many pencils there are in all.

2. Write a word problem that can be represented with the expression 40 ÷ 5. Use the picture to help you.

Sample: There are 40 shopping carts arranged in 5 equal rows. How many shopping carts

are in each row?

I see 40 shopping carts in all.

The shopping carts are in 5 equal rows.

I can ask how many shopping carts are in each row.

REMEMBER

3. Grade 3 students voted for their favorite fruits. The table shows how many students voted for each fruit. Use the data in the table to draw a scaled bar graph.

Type of Fruit	Number of Votes
Apple	25
Orange	15
Strawberry	50
Peach	30

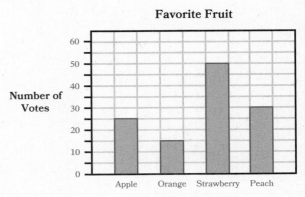

Favorite Fruit

To complete the bar graph, I need to finish the scale and draw the bars.

I need a scale that makes sense for the data in the table. The lowest value is 15. The highest value is 50. I can skip-count by fives. Each tick mark will represent 5 votes. I know I do not have to label every tick mark, so I only label the 10s.

Now I can draw bars for the data. The top of each bar lines up with a number on the scale.

To graph the votes for apple, I draw a bar to 25 units on the scale.

To graph the votes for orange, I draw a bar to 15 units on the scale.

The bars for 15 and 25 stop halfway between labeled tick marks because they are halfway between multiples of 10.

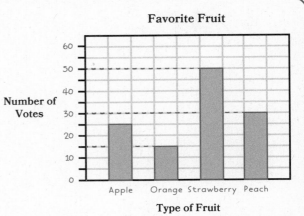

Favorite Fruit

Name _____

1. Write a word problem that can be represented with the expression 5 × 8. Use the picture to help you.

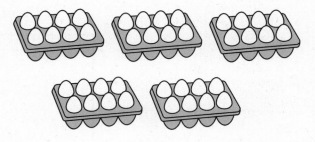

_____

_____

_____

2. Write a word problem that can be represented with the expression 24 ÷ 4. Use the picture to help you.

_____

_____

_____

REMEMBER

3. Students at an elementary school voted for their favorite school lunch. The table shows how many students voted for each lunch. Use the data in the table to complete the scaled bar graph.

School Lunches	Number of Votes
Tacos	150
Chicken Nuggets	200
Hamburger	175
Pizza	225

Favorite School Lunches

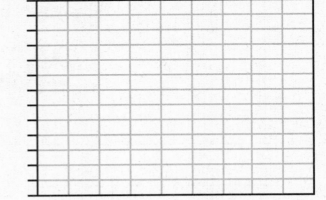

Number of Votes

School Lunch

19

Name

Use the Read–Draw–Write process to solve the problem.

1. Zara buys 9 packs of stickers. Each pack has 8 stickers. Zara gives her friend 37 stickers.

   How many stickers does Zara have left?

   a. Draw to represent the problem. Use a letter to represent each unknown.

   b. Estimate how many stickers are left. Use the questions to help you.

   About how many stickers did Zara buy?

   Sample: $10 \times 8 = 80$

   About how many stickers did Zara give away?

   Sample: 40

   So about how many stickers does Zara have left?

   Sample: $80 - 40 = 40$

   I read the problem. I read again.

   As I reread, I think about what I can draw.

   I draw a tape diagram to represent the 9 packs of stickers and put 8 in each pack. I label the tape diagram with an **m** to represent the unknown, the total amount of stickers.

   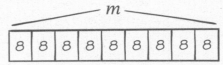

   I draw another tape diagram to represent the second part of the problem. This tape diagram is the same size as the first to show that the total is the same, but this tape diagram shows two parts. It shows the 37 stickers Zara gave away and the unknown stickers that are left. I label this unknown with the letter **n**.

   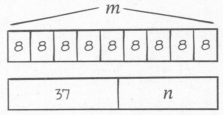

   My drawing can help me estimate how many are left. I think about 10 groups of 8 instead of 9 groups of 8 because I know $10 \times 8 = 80$. Then I round 37 to 40 to make it simpler to subtract. $80 - 40 = 40$.

c. Solve the problem. Write equations and a solution statement.

Sample:

$$9 \times 8 = m$$

$$m = 72$$

$$72 - 37 = n$$

$$n = 35$$

Zara has 35 stickers left.

I multiply 9 and 8 to find the total number of stickers. Zara has 72 stickers in all.

She gave away 37 stickers, so I subtract 37 from 72 to find the number of stickers Zara has left. Zara has 35 stickers left.

I compare the answer to my estimate to see if it is reasonable. 37 is close to 40, so I know my answer is reasonable.

d. How do you know your answer is reasonable?
Use your estimate from part (b) to help you explain.

Sample: My estimate is 40, and my answer is 35.
My answer is reasonable because 35 is close to 40.

REMEMBER

The cafeteria counted how many students chose each type of drink at lunch. The data is shown in the graph.

2. How many fewer students chose chocolate milk than water and white milk combined?

$$90 + 110 = 200$$

$$200 - 160 = 40$$

40 fewer students chose chocolate milk than water and white milk combined.

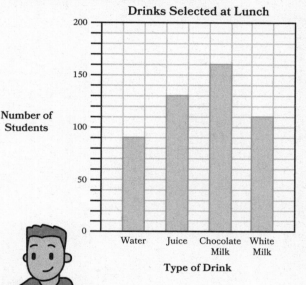

I can use the scaled bar graph to tell me how many students chose each type of drink.

I can draw a line from the top of each bar to the scale and read the number of students who chose each drink. I can label the top of each bar with the number to help me answer the question.

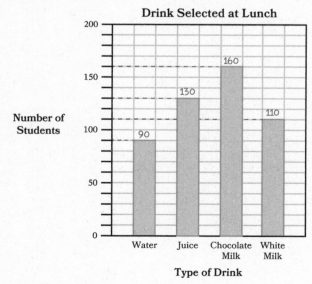

To solve the problem, first I need to know the combined number of students who chose water and who chose white milk. I can add the total number of choices for each drink.

$$90 + 110 = 200$$

Then I subtract the number of choices for chocolate milk from the combined total.

$$200 - 160 = 40$$

19

Name _____

Use the Read–Draw–Write process to solve the problem.

1. Mrs. Smith buys 9 boxes of markers for her classroom. Each box has 6 markers.

   Mrs. Smith uses 28 markers during the school year.

   How many markers are left?

   a. Draw to represent the problem. Use a letter to represent each unknown.

   b. Estimate how many markers are left. Use the questions to help you.
      About how many markers did Mrs. Smith buy?

      About how many markers did Mrs. Smith use?

      So about how many markers are left?

   c. Solve the problem. Write equations and a solution statement.

   d. How do you know your answer is reasonable? Use your estimate from part (b) to help you explain.

REMEMBER

Students counted the number of each type of attraction at a theme park. The data is shown in the graph.

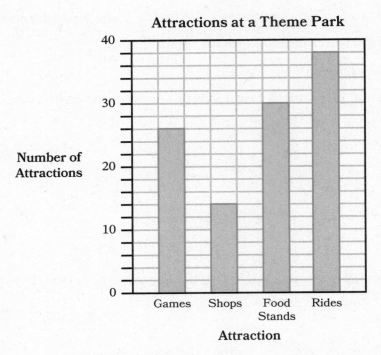

Attractions at a Theme Park

Number of Attractions

Attraction

2. How many fewer food stands are there than games and shops combined?

FAMILY MATH

Multiplication with Multiples of 10 and Further Application of Concepts

Dear Family,

Your student is learning how to multiply by multiples of 10 up to 90: 10, 20, 30, 40, 50, 60, 70, 80, and 90. They use what they know about place value and other strategies to break apart a multiple of 10 into two factors. They apply their multiplication skills to solve two-step word problems and to count more complex groups of objects. The strategies your student is learning now will support their understanding of multiplying with larger numbers.

$$4 \times 2 \text{ ones} = 8 \text{ ones}$$
$$4 \times 2 = 8$$

$$4 \times 2 \text{ tens} = 8 \text{ tens}$$
$$4 \times 20 = 80$$

Writing multiples of ten in unit form shows the connection between multiplying by ones and multiplying by tens.

$$4 \times 20 = 4 \times (2 \times 10)$$
$$= (4 \times 2) \times 10$$
$$= 8 \times 10$$
$$= 80$$

Breaking apart a multiple of 10 into a number times 10 allows for multiplying smaller, known facts first.

At-Home Activities

Multiplying by Tens

Look for items that come packaged in multiples of 10. Encourage your student to use the items to practice multiplying by multiples of 10.

- "One package of pencils comes with 40 pencils. How many pencils would be in 7 packages?"
- "There are 8 bags of oranges on that shelf. Each bag holds about 20 oranges. About how many oranges are on the shelf?"
- "This shoe rack holds 30 shoes. How many shoes will 3 shoe racks hold?"

Tens of Cents

Help your student count a collection of coins. Provide an assortment of nickels, dimes, and quarters, or write 5¢, 10¢, and 25¢ on small pieces of paper to represent coins. Have your student put the coins into groups that have the same value. For example, if the value is 30¢, groups could be made from 3 dimes, 6 nickels, or 1 quarter and 1 nickel. Then ask your student to skip-count or multiply to find the total value of all the coins. Their count may sound like, "3 tens, 6 tens, 9 tens, 12 tens" or "30 cents, 60 cents, 90 cents, 120 cents."

As a challenge, suggest they break the coins into two types of groups, each with a different value. For example, they could organize all the dimes into groups of 20¢ and all the nickels and quarters in groups of 30¢. Then have them find the value of the coins in each type of group and add to find the total value of all the coins.

20

Name _____

Complete the equations. Use the place value charts to help you.

1.

tens	ones
	● ● ●
	● ● ●
	● ● ●
	● ● ●
	● ● ●
	● ● ●

6 × 3 ones = ___18___ ones

6 × __3__ = __18__

2.

tens	ones
● ● ●	
● ● ●	
● ● ●	
● ● ●	
● ● ●	
● ● ●	

6 × 3 tens = ___18___ tens

6 × __30__ = __180__

The place value chart in problem 1 shows 6 rows of 3 ones, which is 18 ones. 6 × 3 ones = 18 ones

The place value chart in problem 2 shows that 6 rows of 3 tens makes 18 tens. 6 × 3 tens = 18 tens

I know that 18 tens is 180. I could skip-count by tens to check.

3. Each bag has 20 bagels. How many bagels are in 4 bags ?

There are 80 bagels in 4 bags.

I can think about this problem in unit form. 20 is 2 tens.

I know 4 × 2 ones = 8 ones, so 4 × 2 tens = 8 tens.

I know that 8 tens is equal to 80.

The multiplication fact stays the same. The unit is different.

REMEMBER

4. Use the picture to complete parts (a) and (b).

a. Write a word problem that can be represented with the expression 20 ÷ 5.

Sample: There are 20 flowers in all.

There are 5 equal groups. How many

flowers are in each group?

I count 20 flowers in all.

There are 5 groups of flowers.

I can ask myself, How many flowers are in each group?

b. Write a word problem that can be represented with the expression 20 ÷ 4.

Sample: There are 20 flowers total. I make groups of flowers with 4 flowers in each group.

How many groups of flowers are there?

There are 20 flowers in all.

There are 4 flowers in each group.

I can ask myself, How many groups of flowers are there?

Name _____

Complete the equations. Use the place value charts to help you.

1.

tens	ones
	••
	••

2×2 ones = _____ ones

$2 \times$ _____ = _____

2.

tens	ones
••	
••	

2×2 tens = _____ tens

$2 \times$ _____ = _____

3.

tens	ones
	••••
	••••
	••••

3×4 ones = _____ ones

$3 \times$ _____ = _____

4.

tens	ones
••••	
••••	
••••	

3×4 tens = _____ tens

$3 \times$ _____ = _____

5. Each classroom can hold 30 students. How many students can 5 classrooms hold?

REMEMBER

6. Use the picture to complete parts (a) and (b).

a. Write a word problem that can be represented with the expression $12 \div 4$.

_____

_____

_____

_____

b. Write a word problem that can be represented with the expression $12 \div 3$.

_____

_____

_____

_____

21

Name _____

Complete the equation. Use the place value chart to help you.

1.

tens	ones

× 10

$(6 \times 4) \times 10 = 24 \times 10$

$= \underline{\quad 240 \quad}$

> I see the place value chart shows 6 rows of 4 ones, which is 24 ones.
>
> 24 ones is multiplied by 10.
>
> Now the place value chart shows 6 rows of 4 tens.
>
> I can think of this as 6 × 4 tens = 24 tens, or as 24 × 10.
>
> 24 tens = 240

2. Place parentheses and fill in the blanks to find each related fact and product.

$$2 \times 60 = 2 \times (6 \times 10)$$
$$= (2 \times 6) \times 10$$
$$= \underline{\quad 12 \quad} \times 10$$
$$= \underline{\quad 120 \quad}$$

> I think about 60 as (6 × 10). There are 2 groups of 6 × 10.
>
> I move the parentheses to group the factors 2 and 6 because that is a more familiar multiplication fact.
>
> I know 2 × 6 = 12.
>
> I can think of 12 × 10 as 12 tens. 12 tens is 120.
>
> So, 2 × 60 = 120.

3. Robin finds 5×30 by thinking about how many tens are in 30. Explain Robin's strategy.

Robin thinks about 30 as 3 tens. She can find 5×3 tens $= 15$ tens. So, $5 \times 30 = 15$ tens, or

$5 \times 30 = 150.$

30 is equal to 3 tens.

I know 5×3 ones $= 15$ ones, so 5×3 tens $= 15$ tens.

15 tens is 150.

REMEMBER

4. Complete the equation to describe the array.

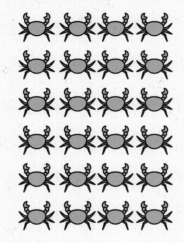

Sample:

____6____ × ____4____ = ____24____

I see 6 rows in the array.

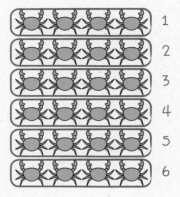

I see 4 in each row.

I can multiply the number of rows by the number in each row.

____6____	×	____4____	=	____24____
Number of rows		Number in each row		Product

Or I can multiply the number in each row by the number of rows.

____4____	×	____6____	=	____24____
Number in each row		Number of rows		Product

Either equation can be used to describe the array.

21

Name

Complete the equation. Use the place value chart to help you.

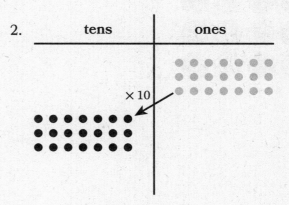

1.

tens	ones

$(4 \times 5) \times 10 = 20 \times 10$

$= \underline{\hspace{1cm}}$

2.

tens	ones

$(3 \times 7) \times 10 = 21 \times 10$

$= \underline{\hspace{1cm}}$

Place parentheses and fill in the blanks to find each related fact and product. The first one is started for you.

3. $4 \times 20 = 4 \times (2 \times 10)$

   $= 4 \times 2 \times 10$

   $= \underline{\hspace{1cm}} \times 10$

   $= \underline{\hspace{1cm}}$

4. $5 \times 70 = 5 \times 7 \times 10$

   $= 5 \times 7 \times 10$

   $= \underline{\hspace{1cm}} \times 10$

   $= \underline{\hspace{1cm}}$

5. Casey finds 60×4 by thinking about how many tens are in 60. Explain Casey's strategy.

_____

_____

_____

REMEMBER

6. Complete the equation to describe the array.

_____ × _____ = _____

Name

Use the Read–Draw–Write process to solve each problem. Use a letter to represent each unknown.

1. Amy wants to buy a new bicycle. She saves $30 each month for 4 months.

   a. How much money has Amy saved?

   $4 \times 30 = m$

   $m = 120$

   Amy has saved $120.

   I read the problem. I read again.

   As I reread, I think about what I can draw.

   I draw a tape diagram to represent the amount of money Amy has saved.

   I know Amy saved the same amount of money each month for 4 months, so I partition my tape diagram into 4 equal parts. She saves $30 each month, so I label each part with 30.

   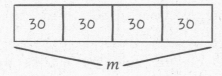

   I can find 4 × 3 tens, or 4 × 30, to find the total amount Amy has saved. I write (4 × 3) × 10. I know that 4 × 3 is 12 and 12 tens is 120.

   b. The bicycle Amy wants costs $145.

   How much more money does Amy need to save to buy the bicycle?

   $145 - 120 = n$

   $n = 25$

   Amy needs to save $25 more.

I read the problem. I read again.

As I reread, I think about what I can draw.

I draw a tape diagram to represent the problem. I label one part with 120 to represent how much money Amy has already saved. I use n to represent how much money Amy still needs. I label the total as 145 to represent the cost of the bicycle.

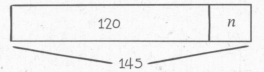

I can subtract 120 from the total cost of the bicycle to find how much more money Amy needs to save.

---

2. Robin has 6 totes of books. Each tote of books has 50 books.

　　a. How many books does Robin have?

　　　$6 \times 50 = r$

　　　Robin has 300 books.

　　b. Robin already had 79 books. How many books does Robin have now?

　　　$300 + 79 = n$

　　　Robin has 379 books now.

I read the problem. I read again.

As I reread, I think about what I can draw.

I draw a tape diagram to represent the number of books Robin has. There are 6 totes of books, so I partition my tape diagram into 6 equal parts.

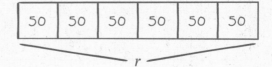

I can find 6 × 5 tens, or 6 × 50, to find the total number of books. I write (6 × 5) × 10. I know that 6 × 5 is 30 and 30 tens is 300.

I can add 79 books to the 300 books from the first part of the problem to find the total number of books.

3. Carla sells 47 tomato plants and 23 bean plants. She earns 3 dollars for each plant she sells.

   What is the total number of dollars Carla earns for selling the tomato plants and bean plants?

   $47 + 23 = p$

   $70 \times 3 = d$

   The total number of
   dollars Carla earns
   for selling the tomato
   plants and bean plants is 210.

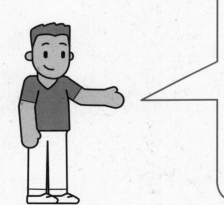

I read the problem. I read again.

As I reread, I think about what I can draw.

I can draw a tape diagram to show the 47 tomato plants and the 23 bean plants. These are the parts. The total is unknown. I label the total with the letter p.

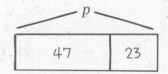

Carla sells 70 plants in all. She earns 3 dollars for each plant.

I can find 70×3 to find the total number of dollars Carla earns.

I write $(7 \times 3) \times 10$. I know that 7×3 is 21 and 21 tens is 210.

REMEMBER

4. Use the picture to write a word problem that can be represented with the expression 2×8.

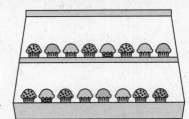

Sample: A display case has 2 shelves. There are 8 muffins on each

shus shelf. How many muffins are in the display case?

_____

_____

I look at the picture.

I see a display case with 2 rows of muffins.

Each row has 8 muffins.

The expression 2 × 8 represents the total number of muffins.

I can ask a question about how many muffins are in the display case.

22

Name _____

Use the Read–Draw–Write process to solve each problem. Use a letter to represent each unknown.

1. There are 20 apple trees in a garden. Luke picks 6 apples from each tree.

   a. What is the total number of apples Luke picks?

   b. Eva picks 165 apples from the apple trees.

      How many more apples does Eva pick than Luke?

2. Mr. Endo buys 9 boxes of colored pencils. Each box has 30 colored pencils.

   a. How many colored pencils does Mr. Endo buy?

   b. Mr. Endo already had 144 colored pencils. How many colored pencils does Mr. Endo have now?

3. Gabe builds 26 tables and 34 chairs. He takes 4 hours to build each table or chair.

   What is the total number of hours Gabe takes to build the tables and chairs?

REMEMBER

4. Use the picture to write a word problem that can be represented with the expression 3 × 9.

_____

_____

_____

_____

23

Name _____

Find the value of each unknown.

1. $11 \times 6 = h$

   $h = 66$

2. $12 \times 7 = r$

   $r = 84$

> I can think of 11 sixes as 10 sixes and 1 six.
>
> I know $10 \times 6 = 60$ and $1 \times 6 = 6$.
>
> $60 + 6 = 66$, so $11 \times 6 = 66$.

> I can think of 12 sevens as 10 sevens and 2 sevens.
>
> I know $10 \times 7 = 70$ and $2 \times 7 = 14$.
>
> $70 + 14 = 84$, so $12 \times 7 = 84$.

3. Ivan draws a tape diagram to represent 12×4. Fill in the blanks to find the total.

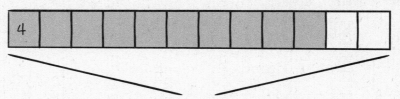

12 fours = 10 fours + 2 fours

$$= \underline{40} + 8$$

$$= \underline{48}$$

> 10 fours is equal to 40, and 2 fours is equal to 8.
>
> I can add 40 and 8 to find 12×4.
>
> I can skip-count by fours to check: 4, 8, 12, 16, 20, 24, 28, 32, 36, 40, 44, 48.

Use the Read–Draw–Write process to solve each problem.

4. 8 baskets have 11 oranges each. How many oranges are there?

$8 \times 11 = 88$

There are 88 oranges.

> I read the problem. I read again.
>
> As I reread, I think about what I can draw.
>
> I draw a tape diagram with 8 equal parts to represent the number of baskets. Each part has 11 oranges. I use n for the unknown total number of oranges.
>
> | 11 | 11 | 11 | 11 | 11 | 11 | 11 | 11 |
>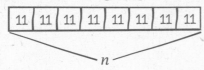
> n
>
> I can multiply 8 and 11 to find the total number of oranges.

5. How many bagels are in 7 dozen?

$7 \times 12 = 84$

There are 84 bagels in 7 dozen.

> I read the problem. I read again.
>
> As I reread, I think about what I can draw.
>
> I draw a number bond to help me. I know that 1 dozen is equal to 12. I can multiply 12 and 7 to find the number of bagels.
>
> $$7 \times 12 = (7 \times 10) + (7 \times 2)$$
> $$10 \quad 2$$
> $$= 70 + 14$$
> $$= 84$$
>
> I break apart 12 into 10 and 2 to multiply simpler facts.
>
> I multiply 7 and 10 to get 70, and I multiply 7 and 2 to get 14. I add the products together to get 84. So, $7 \times 12 = 84$.

REMEMBER

6. Circle the two equations that show the commutative property of multiplication.

$$3 \times 6 = 6 + 6 + 6$$ 　　　 $$\boxed{3 \times 8 = 8 \times 3}$$

$$\boxed{2 \times 5 = 5 \times 2}$$ 　　　 $$2 \times 6 = 3 \times 4$$

> The **commutative property of multiplication** says I can change the order of the factors and get the same product.

> $3 \times 6 = 6 + 6 + 6$
>
> This equation has an addition expression on one side. Factors are the numbers being multiplied.
>
> So, this equation does not show the commutative property.

> $3 \times 8 = 8 \times 3$
>
> Both sides of this equation have the same factors, 3 and 8. The factors are in a different order on each side of the equal sign.
>
> This equation shows the commutative property.

> $2 \times 5 = 5 \times 2$
>
> Both sides of this equation have the same factors, 2 and 5. The factors are in a different order on each side of the equal sign.
>
> This equation also shows the commutative property.

> $2 \times 6 = 3 \times 4$
>
> Each side of this equation has different factors.
>
> This equation does not show the commutative property.

23

Name _____

Find the value of each unknown.

1. $4 \times 11 = n$

2. $12 \times 6 = m$

3. Oka draws a tape diagram to represent 12×9. Fill in the blanks to find the total.

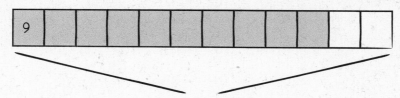

12 nines = 10 nines + 2 nines

= _____ + 18

= _____

Use the Read–Draw–Write process to solve each problem.

4. 6 shelves have 11 books each. How many books are there?

5. How many eggs are in 4 dozen?

REMEMBER

6. Circle the two equations that show the commutative property of multiplication.

$$4 \times 6 = 6 + 6 + 6 + 6 \qquad 2 \times 6 = 3 \times 4$$

$$3 \times 6 = 6 \times 3 \qquad 6 \times 7 = 7 \times 6$$

Name

REMEMBER

Use the Read–Draw–Write process to solve the problem.

1. James has 19 cents more than Luke.

    James has 2 quarters, 2 dimes, 1 nickel, and 1 penny.

    How much money does Luke have?

    Sample:

    $50 + 20 + 5 + 1 = 76$

    $76 - 19 = 57$

    Luke has 57 cents.

I read the problem. I read again.

As I reread, I think about what I can draw.

I know James has 2 quarters, 2 dimes, 1 nickel, and 1 penny.

I can draw circles to represent each type of coin and label their values.

I can add the values of the coins to find how much money James has.

I know James has more money than Luke. That means Luke has less money than James.

I can find how much money Luke has by subtracting 19 cents from James's total. I use the open number line to record my thinking.

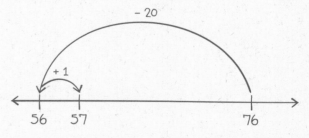

24

Name _____

REMEMBER

Use the Read–Draw–Write process to solve the problem.

1. Liz has 27 cents more than Zara.

   Liz has 3 quarters, 2 nickels, and 5 pennies.

   How much money does Zara have?

25

Name _____

Use the Read–Draw–Write process to solve each problem.

Ray invites 84 people to a backyard movie party.

1. Ray buys 7 bags of chips. What is the total cost of the chips?

   $7 \times 6 = 42$

   The total cost of the chips is $42.

Item	Cost of Each
Candy Bars (Box of 8)	$8
Drinks	$11
Bag of Chips	$6
Jumbo Popcorn Bag	$10

I read the problem. I read again.

As I reread, I think about what I can draw.

I draw and label a tape diagram to represent the total cost of the chips.

There are 7 bags of chips, so I partition my tape diagram into 7 equal parts. Each bag of chips costs $6.

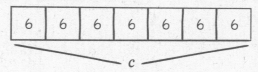

I can multiply 7 and 6 to find the total cost of the chips.

2. Ray buys 12 boxes of candy bars.

   a. Does he buy enough candy for everyone to have 1 candy bar? Explain how you know.

   $12 \times 8 = 96$

   Yes, 84 people are invited to the movie party, and Ray buys 96 candy bars.

   b. What is the total cost of the candy bars?

   $12 \times 8 = 96$

   The total cost of the candy bars is $96.

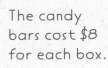

> I read the problem. I read again. As I reread, I think about what I can draw.
>
> I draw a tape diagram to represent the 12 boxes of candy bars. There are 8 candy bars in each box.
>
>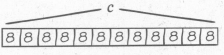
>
> I can multiply 12 and 8 to find the total number of candy bars Ray buys. I know $10 \times 8 = 80$ and $2 \times 8 = 16$. I add 80 and 16 to find the total number of candy bars.
>
> Ray buys 96 candy bars. 96 is greater than 84, so he buys enough for everyone to have 1 candy bar.

> The candy bars cost $8 for each box.
>
> I can use my tape diagram from part (a) to help find 12 eights.

3. Ray spends $159 on popcorn and drinks. $99 is spent on drinks. How many bags of popcorn does he buy?

   $159 - 99 = 60$

   $60 \div 10 = 6$

   Ray buys 6 jumbo bags of popcorn.

> I read the problem. I read again.
>
> As I reread, I think about what I can draw.
>
> I draw a tape diagram to represent the amount Ray spends on popcorn and drinks.
>
>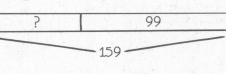
>
> The tape diagram helps me see that I can subtract the cost of the drinks from $159 to find the total amount Ray spends on popcorn.
>
> Then I can divide the total Ray spends on popcorn by the cost of each bag of popcorn to find the number of bags of popcorn Ray buys.
>
>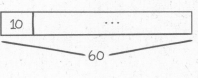

4. What is the total cost of the food and drinks that Ray buys?

$42 + 96 + 159 = 297$

The total cost of the food and drinks that Ray buys is $297.

Ray spends $42 on chips, $96 on candy, and $159 on popcorn and drinks. I can add all these amounts to find the total cost of the food and drinks.

REMEMBER

5. Complete the equations to find $42 \div 6$.

$42 \div 6 = (30 \div 6) + (\underline{\ \ 12\ \ } \div 6)$

$= \underline{\ \ 5\ \ } + \underline{\ \ 2\ \ }$

$= \underline{\ \ 7\ \ }$

I can use a number bond to break apart 42 into smaller parts, which I can divide by 6 by using facts I know.

I can break apart the division problems by using the smaller parts. I divide each part by 6 and add the quotients.

$30 \div 6 = 5$ $12 \div 6 = 2$

$5 + 2 = 7$

✏ 25

Name _____

Use the Read–Draw–Write process to solve each problem.

Mia invites 60 people to a pool party.

1. Mia buys 3 snack trays. What is the total cost of the snack trays?

Item	Cost of Each
Drinks (Pack of 12)	$9
Party Sub	$35
Watermelon	$8
Snack Tray	$10

2. Mia buys 6 packs of drinks.

   a. Does she buy enough drinks for everyone to have 1 drink? Explain how you know.

   b. What is the total cost of the drinks?

3. Mia spends $83 on a party sub and watermelons. How many watermelons does she buy?

4. What is the total cost of the food and drinks that Mia buys?

REMEMBER

5. Complete the equations to find $56 \div 8$.

$56 \div 8 = (40 \div 8) + (\underline{\hspace{1cm}} \div 8)$

$= \underline{\hspace{1cm}} + \underline{\hspace{1cm}}$

$= \underline{\hspace{1cm}}$

Acknowledgments

Kelly Alsup, Lisa Babcock, Cathy Caldwell, Mary Christensen-Cooper, Cheri DeBusk, Jill Diniz, Melissa Elias, Janice Fan, Scott Farrar, Krysta Gibbs, Julie Grove, Karen Hall, Eddie Hampton, Tiffany Hill, Robert Hollister, Rachel Hylton, Travis Jones, Liz Krisher, Courtney Lowe, Bobbe Maier, Ben McCarty, Maureen McNamara Jones, Cristina Metcalf, Melissa Mink, Richard Monke, Bruce Myers, Marya Myers, Geoff Patterson, Victoria Peacock, Marlene Pineda, Elizabeth Re, Meri Robie-Craven, Jade Sanders, Deborah Schluben, Colleen Sheeron-Laurie, Jessica Sims, Theresa Streeter, Mary Swanson, James Tanton, Julia Tessler, Saffron VanGalder, Jackie Wolford, Jim Wright, Jill Zintsmaster

Trevor Barnes, Brianna Bemel, Adam Cardais, Christina Cooper, Natasha Curtis, Jessica Dahl, Brandon Dawley, Delsena Draper, Sandy Engelman, Tamara Estrada, Soudea Forbes, Jen Forbus, Reba Frederics, Liz Gabbard, Diana Ghazzawi, Lisa Giddens-White, Laurie Gonsoulin, Nathan Hall, Cassie Hart, Marcela Hernandez, Rachel Hirsh, Abbi Hoerst, Libby Howard, Amy Kanjuka, Ashley Kelley, Lisa King, Sarah Kopec, Drew Krepp, Crystal Love, Maya Márquez, Siena Mazero, Cindy Medici, Ivonne Mercado, Sandra Mercado, Brian Methe, Patricia Mickelberry, Mary-Lise Nazaire, Corinne Newbegin, Max Oosterbaan, Tamara Otto, Christine Palmtag, Andy Peterson, Lizette Porras, Karen Rollhauser, Neela Roy, Gina Schenck, Amy Schoon, Aaron Shields, Leigh Sterten, Mary Sudul, Lisa Sweeney, Samuel Weyand, Dave White, Charmaine Whitman, Nicole Williams, Glenda Wisenburn-Burke, Howard Yaffe

Credits

For a complete list of credits, visit http://eurmath.link/media-credits